THE SOUND ATLAS

The Sound Atlas

A Guide to Strange Sounds across Landscapes and Imagination

Michaela Vieser
and
Isaac Yuen

REAKTION BOOKS

Published by
REAKTION BOOKS LTD
2–4 Sebastian Street
London EC1V 0HE, UK
www.reaktionbooks.co.uk

First published in English 2025
Copyright © Michaela Vieser and Isaac Yuen 2025
First published in German by Knesebeck Verlag 2023

All rights reserved

EU GPSR Authorised Representative
Logos Europe, 9 rue Nicolas Poussin, 17000, La Rochelle, France
email: contact@logoseurope.eu

No part of this publication may be reproduced, stored in a retrieval system or transmitted, in any form or by any means, electronic, mechanical, photocopying, recording or otherwise, without the prior permission of the publishers. No part of this publication may be used or reproduced in any manner for the purpose of training artificial intelligence technologies or systems.

Printed in Great Britain by TJ Books, Padstow, Cornwall

A catalogue record for this book is available from the British Library

ISBN 978 1 83639 110 4

Contents

	Preface	9
1	Sounds of the Universe: In Search of Harmony and Resonance	13
2	The Shape of Sound: The Singing Pillars of Hampi	17
3	An Arms Race 50 Million Years in the Making: The Shadow War of Moths and Bats	21
4	Caves, Sounds and the Human Imagination	25
5	Reconstruction of the Paleosonic: Bringing Back the Sounds of Past Life	29
6	Calling Out Across Far Distances: Sound Messages in Air and Water	32
7	Reverberations Across Time and Fate: The Oracle at Dodona	38
8	Word on the Wind: The Curious History of the Aeolian Harp	41
9	To Be Defined By Sound: Regarding the Cicada	45
10	*Gagaku*: A Music that Crosses Space and Time	49
11	Sounds of the Cryosphere	54
12	The Stories and Secrets of Singing Sands	58

13	The Humility Pipe	62
14	The Untameable Ritual of Keening	67
15	Death of a Sound and Age: Pining for the Foghorn	71
16	A Voice, an Echo, Silence	75
17	Sounds of the Atomic Age	79
18	Sounds from the Deepest Artificial Point on Earth	83
19	The Humming Fields and Meadows of the Altai	88
20	Otherworldly Ordinary: The Found Sounds of the Fantastic	92
21	Pay Attention: On the Sounds of the In-Between	95
22	The Not-So-Secret Love Lives of Fish	100
23	Voices Within and Without	103
24	Listening to Traces	107
25	Explaining the Inexplicable: The Taos Hum	112
26	One Square Inch of Silence	116
27	Acoustics Accidental and Incidental	120
28	Sound Over Sight as Sense: The World According to Whales	124
29	Geophony, Biophony, Anthrophony: The Three Sound Types Beneath the Seven Seas	128
30	'That's Not What It's Supposed to Sound Like': Bizarre Bird Calls from All Seven Continents	132
31	Sensing the Sound of a Landscape through Rock and Stone	135
32	The Body Fields and the Works of Jacob Kirkegaard	138
33	ASLSP: As Slow As Possible	142

34 From Wax and Glass, Music and Voices: The Past,
   Present and Future of Sound Recording   *147*

35 The Sound of Manipulation: Sonic Warfare
   and Propaganda   *151*

36 Humanity's Message for Whom? The Voyager
   Golden Record   *156*

BIBLIOGRAPHY   *161*

ACKNOWLEDGEMENTS   *175*

Preface

> I have become a listener.
> BERNIE KRAUSE, *bioacoustician*

Ask a friend: where on Earth is most beautiful? You may hear about a white sand beach, a windswept valley, a medieval castle town. While these places may be interesting, they most likely will not reveal much about your friend's inner landscapes. But ask them about which soundscape has moved them, and you may get something more illuminating. One friend we did this little exercise with took us on an unexpected journey:

> My boys were still young, I just wanted to get them out of the house, so I suggested going into the forest. It was one of those November afternoons where the clouds hang low as a kind of thick fog, with droplets of water clinging on to each other. Suddenly a wind started to blow through the tip of the fir trees. It was as if someone was playing a harp. And I called out to my boys: 'The trees are singing for us.' And my boys nodded their heads. They could hear and feel it too. I have never told

this to anyone. This sound, this moment, that I now remember.

The phenomena of sound may be fleeting and evanescent, but the memory of it can open a profound window into the soul.

Sound is energy. Its waves penetrate our bodies and unsettle our cores. We can close our eyes to sights we do not wish to see, but we are less able to shield our bodies against the effects of sound. While ultrasonic or infrasonic frequencies cannot be listened to with our ears, they nevertheless travel through our tissues and jolt our cells. Read the chapter on the world's largest organs to find out how the use of certain tones can shape and manipulate our emotions. Learn how hard it is to escape sound unless you are a moth, evolved to elude detection by hungry bats. Or mellow to the fact that there are waves of energy rippling across the universe from the aftermath of the Big Bang, reaching us only now, billions of years later.

So much of our knowledge base is visual, revolving around colours, shapes and patterns, so much so that we can often lock ourselves into this one-sense-perspective: we see, watch, sort, block out and imagine with our mind's eye. But our bodies experience and process so much more. Is it possible that our over-reliance on sight has made us oblivious to the perception of other worlds? What if we are not recognizing realities that are right in front of us?

A sonic revolution is under way. The expanding field of acoustics is enabling us to tune in to a multitude of invisible worlds. Scientists are transforming energy transmissions from

Preface

distant galaxies into audible harmonics. Sound artists use vibration sensors to probe into the deepest parts of the world. Sophisticated hydrophones are revealing the vibrant chatter in our rivers and oceans. But with this ability to eavesdrop into newfound realms comes a responsibility not only to understand, but to safeguard them – often from ourselves.

With the onset of the Anthropocene and the age of climate turmoil, our species has unintentionally shaped new soundscapes while obliterating others. Humans have introduced noise and chaos where once there was silence. We have extinguished what used to have rhythm, melody and harmony. While researching this book, we felt many of these losses reverberate and echo through us.

This book is an attempt to tell the stories behind sounds that might otherwise seem mundane or ordinary. It is an exercise to delve deeper into the layers of a multitude of acoustic experiences. One way to read it is to heed its title, pinging from location to location like an atlas. Another way is to follow its structure, laid out chronologically, stretching from the first explosions flung across the universe to transmissions humanity has cast out into the far future. In-between, the cries of beasts living and extinct, strange hums from mysterious sources, holy sounds and timeless sounds, stretches of silence and ambience, the songs of earth and wind and ice. We hope this book will help you embark on your own journey of sonic wonders found on this riotous planet and beyond.

1

Sounds of the Universe: In Search of Harmony and Resonance

> Holmdel, New Jersey, USA | Perseus Galactic Cluster, 240 million light years away | The TRAPPIST-1 system, 39 light years away in the constellation of Aquarius

In 1964 radio astronomers Arno Penzias and Robert Wilson repurposed a commercial antenna in Holmdel, New Jersey, to conduct research into the spaces between galaxies.

To their dismay, a constant, low-level noise cropped up in their measurements. Wherever they pointed their instrument – galaxy clusters, supernova remnants, New York City – the same background hiss. The suspected culprits: a flock of pigeons roosting inside the helium-cooled, carefully calibrated antenna. Initial attempts at relocation proved unsuccessful, leading to resolution via technician with shotgun – neither Penzias nor Wilson has ever claimed responsibility for giving the order.

The noise persisted, radiating from every region of sky regardless of time of day. After months of troubleshooting, the researchers turned to Princeton physicist Robert Dicke. Together, they realized that they had been picking up the left-over signature from a massive explosion 13.8 billion years ago. For their accidental discovery of cosmic microwave background radiation, the afterglow that verified the Big Bang Theory, Penzias and Wilson were awarded the Nobel Prize in Physics in 1978.

No need to visit the now national historical landmark in New Jersey to pick up the oldest sound of all. Simply tune in to an empty channel on any radio or old analogue television – amid the static, a residual trace from the birth of the universe. Wilson still keeps a recording of the signal on his mobile phone. Some have compared it to the roar of the ocean, overwhelming, unfathomable. Others perceive it as white noise capable of drowning out all worries, a lullaby for drifting off to a sweeter and shorter form of oblivion.

Amid this din, what other objects are sounding out across the depths of the cosmos? Most appear not to be meant for human ears, like the sound waves generated from a supermassive black hole at the centre of the Perseus galactic cluster – a full 57 octaves lower than middle C and a million billion times below our range of hearing. A constant B flat for 2.5 billion years – perhaps it is a good thing that sound cannot travel across a vacuum.

To hear across the void we must rely on indirect means, converting other forms of emissions – light, radio, X-rays, gravity – into something we can discern, and derive pattern and meaning from. Sonification is the process of transforming data into audible registers. The two mandates of the group SYSTEM Sounds are to:

1. Translate the rhythm and harmony of the cosmos into music and sound.
2. Inspire, educate and make astronomy accessible to the visually impaired.

Developed in collaboration with NASA's Chandra X-Ray Center, their 'Universe of Sound' project transports listeners to places and phenomena they can scarcely imagine through the sonic interpretations of telescope images. Listen to the interplay of sounds mapped to four elements scattered by the exploded star Cassiopeia A. Hear the arrangement of brass, strings and woodwind derived from the whirling neutron star at the heart of the Crab Nebula. Absorb the eerie trills generated by transforming light and X-rays from the Pillars of Creation. Extracting aesthetic qualities from the cold, atonal emanations of space – sonification is as much an art as a science, a means for both comprehension and appreciation.

Perhaps sonic beauty can be discerned within movement itself. So believed Pythagoras of Samos, player of lutes, deviser of theorems, proposer of *musica universalis*, also known as 'the harmonies of the spheres', the concept that even the distances between stars and planets can be translated into music. 'There is geometry in the humming of the strings. There is music in the spacing of the spheres.'

A present-day Pythagoras of Earth would be disappointed to learn that our solar system harbours no such melodies, at least to human ears; the orbits of our eight planets stretch too far to make for a pleasant listening experience. But a potential Pythagoras of TRAPPIST-1 would be delighted to discover that the seven planets of his solar system are locked in orbital resonance. For each two orbits made by the outermost planet, the next one in orbits thrice; the next, four times; then six, nine, fifteen, twenty-four – ratios simple enough for SYSTEM Sounds to transform into music. One can insert additional flourishes while playing with the free sonification model on their website: a drum hit every time a planet passes another to denote the tug of gravity. An extra beat to account for the dwarf star's fluctuating brightness. The result: something that almost borders on a piece being composed, a stellar orchestra coming into tune.

Scientists have discovered more than 5,000 exoplanets to date; the number roughly doubles every two years. Perhaps in the not-too-distant future, we will chance upon a system that has been resonating in perfect harmony, divining the music in motion exactly as dreamt up by ancient philosophers gazing up at the heavens, searching for meaning so long ago.

2

The Shape of Sound: The Singing Pillars of Hampi

Vittala Temple, Hampi, India | Delhi National Museum, India | CERN, Geneva, Switzerland

The rocks in Hampi are old, even for stones. No volcano spat them out, no tectonic shift heaved them high: the granite boulders shaping this southern Indian landscape are part of a so-called inselberg, the crust of a continent that existed 3.6 billion years ago. While everything around has been withered away by wind and weather, they have kept their hold. Geologists get sweaty when they tell this story of deep time.

'The size of this city I do not write here, because it cannot all be seen from any one spot, but I climbed a hill whence I could see a great part of it . . . What I saw from thence seemed to me as large as Rome, and very beautiful to the sight,' wrote the Portuguese missionary Domingo Paes upon his visit to the city of Vijayanagar, modern-day Hampi, in the sixteenth century. 'The people in this city are countless in number, so much so that I do not wish to write it down for fear it should be thought fabulous.' He lists the many groves of trees within its precinct, the conduits of water which flow into the midst of its houses, exalts at the achievement of the city having food for everyone and waxes about its lively bazaars. In those days, only Beijing was bigger.

Empires do not last. What once was Vijayanagar now lies scattered over an area of 41 square kilometres (16 sq. mi.), still mythical and magic, even in bright daylight. But when the rays of the setting sun tint the rocks and the city's remains with apricot hues, it is as if the stones would start to sing.

Which is exactly what they do. At the ruins of Hampi's fourteenth-century Vittala Temple, 56 musical granite pillars hold up a roof. When tapped with a light flick of the fingers some resonate like cymbals, drums or conch shells. Some oscillate like flutes and harps. Others resemble the Jaltarang, a classical Indian instrument which uses different sized cups filled with water, each giving off a different note when struck. While some of the pillars resonate with the seven Indian musical notes of Saptaswara, others follow the five tone system.

It is very likely that someone played classical ragas on this resounding architecture, which have the ability to 'colour the mind' of an audience. In Hinduism, such songs were thought to have their own natural existence within this world. The musician was not to invent their melodies and tunes, but to discover and explore the song's qualities. They could open up a portal to liberate mind and conscience, enabling the listener to achieve *moksha*: to once and for all leave the cycle of rebirth. Hence music was not considered a commodity but a spiritual path, and the musician its wayfinder.

It remains a mystery as to how the scientists, alchemists and architects of the old empire constructed the musical pillars and made them work like synthesizers, replicating already existing sounds or creating utterly new ones. When modern-day sound engineers examined the site, they discovered that some of the pillars when struck reverberated at frequencies lower than the human ear can hear. Such waves would then interact with other pillars and cause them to tremble in return, producing a resonance of interplay and a soundscape that came alive by its own makings.

The *Ryōjin hishō* (Songs that Make the Dust Dance) is a collection of Japanese songs from the twelfth century. The patterns that the dust leaves behind when it settles after being stirred up by music are referred to by acoustics researchers as cymatics, a phenomenon that would be examined by Galileo Galilei during the Renaissance and by Robert Hooke during the Age of Reason. Only recently have artists like Björk integrated cymatics into their music, visualizing what patterns emerge when sound waves move lightweight matter.

The ceiling above the various pillars in Hampi's Vittala Temple is ornamented with different flower- or mandala-like stone carvings. These may serve as visual representations of the sound waves created within this space. Each room lending to a different experience, each room providing the perfect soil for one particular sound flower. A still of the movement of time, matter and energy.

It is said that before his marriage to Parvati, the god Shiva meditated in Hampi. When the couple finally united, the gods rained *aurum*, or gold particles, over the landscape, made visible every day since in the morning or evening glow. Every sunrise and sunset a new cymatic, a new blossoming of the song of the world.

In Hampi, the golden-hued boulders will endure until rain and wind wear them down, transforming them at last to dust in dance. If only the vyalas, the stone statues guarding the entrance to the musical pillars, could speak. Part lion, griffin, tiger or elephant in combat, they symbolize the triumph of spirit over matter.

One of the most striking bronze figures ever created is the dancing Shiva Nataraja, now at the Delhi National Museum, dating from the twelfth century. Depicted is the lord of dance as he moves across the universe, dancing the Ananda Tandava, the dance of bliss. Waves emerge from his head, gathering into a halo, while the vibrations of his drum are said to sort out matter, energy and sound. A replica of this statue was chosen as a landmark by CERN in Switzerland, the world's largest particle laboratory. Shiva, it is said, will dance until the end of days and until a new beginning, guided by the most ancient of sounds.

3

An Arms Race 50 Million Years in the Making: The Shadow War of Moths and Bats

Bracken Cave, Texas, USA | Southwestern Research Station (SWRS), Arizona, USA

As the sun sets on a summer day, an epoch-spanning war in the night sky resumes. We may not have been privy to this nocturnal conflict if not for American neuroscientist Robert Galambos, who coined the term 'echolocation' back in 1944 to describe how bats hunt by emitting ultrasonic calls and processing reflections. This discovery came at a time of another conflict seemingly without end, when nations began to realize the strategic value of perceiving objects that lie beyond the measure of sight.

The first demonstration of the use of radar for detecting aircraft occurred in the English market town of Daventry, on 26 February 1935. The biosonar used by bats has been developed over 50 million years. Current evidence in the fossil record supports the theory that powered flight came before blindsight, suggesting bat ancestors were flying before navigating the world solely on sound. Flying is a calorie-intensive endeavour, and echolocation might have allowed bats to exploit a night sky that abounded with food.

Moths, in particular, prove the ideal morsel for many insect-eating bat species around the world. Millions of Mexican free-tailed bats fly out of Bracken Cave in Texas from spring to autumn every year to feast on cotton bollworm and army cutworm moths. Capable of flying at an altitude of up to 3 kilometres (2 mi.) with a top horizontal speed of over 160 kilometres per hour (99 mph), their performance profiles far outstrip any swift or swallow. Unloved, unlauded and at home in the dark, bats are the phantom aces of the animal kingdom.

What can meagre moths do against such sophisticated hunters? Quite a lot, it turns out. In nature, as in protracted warfare, measures beget countermeasures. Over 100,000 species of moths exist around the world; many have developed defences against their deadliest foes. Some have evolved sensory apparatus to detect the clicks and chirps of echolocating bats, affording them time to dive and swoop in a night-time dogfight. The greater wax moth possesses arguably the most sensitive ears in existence, capable of detecting frequencies of up to 300 kHz – the limit of human hearing caps off at 20 kHz. It is able to pick up even the highest squeaks of the Percival's trident bat, a bat record at 212 kHz.

Other moths, being completely deaf, have opted for another strategy: stealth. Biologist Marc Holderied from the University of Bristol found that the furry thorax of the Suraka silk moth absorbs incoming ultrasound and returns unreadable signatures back to hunting bats. Holderied also found that the wings of the Chinese tussar moth comprise a metastructure of tuned scales that uses resonance to swallow up at least three octaves of sound: an intricate, impossibly thin acoustic cloak that leaves bats hungry and material engineers scratching their heads.

Should advance detection and nigh invisibility prove insufficient, some moths even utilize sonic shadows and active jamming. The American moon moth has evolved trailing hindwings that create sensor ghosts and disrupt the ability of big brown bats to hit vital body parts. The Grote's tiger moth is capable of emitting a burst of clicks to throw off the timing of incoming mouse-eared bats, turning certain strikes into near misses. 'Bats captured ten times more silenced moths as clicking moths,' noted behavioural ecologist Aaron Corcoran while working at the Southwestern Research Station in Arizona. But only a narrow window exists for effective disruption. Moth clicks must occur within a 1–2 millisecond time frame in order to interfere with the bat's tracking system: a razor-thin margin between becoming dinner and fluttering off to live another day.

Some moth broadcasts serve less as interference and more as explicit warning: *Mess with me and you'll be sorry*. Upon being targeted, Martin's lichen moths sound out their noxious natures to would-be predators, which have come to associate their calls with a terrible taste. According to Nicolas Dowdy, Head of Zoology at the Milwaukee Public Museum, these moths appear to be more 'nonchalant' during their flights and less likely to conduct evasive manoeuvres, a testament to the degree of confidence they have in their chemical defences, developed through the ingestion of poisonous plants in earlier caterpillar phases.

This strategy, called acoustic aposematism, can also provide umbrella protection for other perfectly palatable copycats, should they be able to mimic the call. Thus the incentive to become an imposter is high, which can then dilute the overall warning message to the detriment of all. This phenomenon, called Batesian mimicry, means that foul-tasting moths have to contend with not only their arch-nemeses, but their delicious-tasting cheating kin acting out of self-interest and bad faith.

So the sounds shift and new strategies emerge. Moths evolve ever more complex ways to counter bat aggression. Bats evolve ever-increasing means to overcome moth defences. There is only adaptation or extinction.

Some moths have chosen to flee from bats and shadows by escaping into the day, into a realm of colour and light. But in attempting to abandon one cycle they have only found themselves trapped within another. For there is another class of foes awaiting as the sun rises – the birds. In nature, there is no real escaping the arms race between predator and prey. There is only engaging with another one anew.

4

Caves, Sounds and the Human Imagination

The Pertosa-Auletta Caves, Italy | The Cave of Altamira and Palaeolithic Cave Art of Northern Spain

The journey into the caves of Pertosa begins at the Hall of Wonders. Board a boat hand-towed by guides to cross the underground river. On the other side you will hear the roar of the black waterfall, enter caverns named the Belvedere and the Great Hall, behold subterranean features 34 million years in the making.

Open to tourists since 1932, the acoustic properties of this natural auditorium have not gone unnoticed by artists and auteurs. In recent years, local theatre groups have regaled visitors with performances of Dante's *Inferno* with the underground setting serving as the ten circles of hell, staged sagas of Ulysses' descent into Hades in search of the soothsayer Tiresias. Italian *giallo* master Dario Argento shot scenes for his 1998 rendition of *The Phantom of the Opera* here. This is not surprising. Arguably no other place on Earth is more suited to serve as lair for an unrequited soul with a flair for the dramatic than a site with chamber names like Sala del trono, or the Throne Room, within which one can play the organ in angst and solitude; or Sala dell'unione, the Union Hall, where stalactite and stalagmite yearn to reach one another, seeking on some distant day to kiss and merge as one.

Sheltered from the outer world of light and change, caves served as incubators for the human impulse to manifest innermost desire into reality. Throughout prehistory, people bore their souls into these dark recesses, driven to impart elemental expressions upon something beyond themselves. Thus aurochs and cave lions rendered within the caves of Chauvet in southern France. Thus a trio of Sulawesi pigs on the walls of Leang Tedongnge in Indonesia. Thus hundreds of stencilled handprints gave name to the cavern known today as Ceuva de las Manos in Argentina.

But while lines in ochre and charcoal can endure across the ages, the aural traces that accompanied these depictions are long gone, diffused into porous limestone. What were the sounds around the creation and celebration of these artistic endeavours? What insights into the minds of early modern humans can be gleaned from the silent scapes they left behind?

Archaeoacoustics explores the connection between people and sound throughout history. It is concerned with that elusive phenomenon that leaves no trace but unites many a tradition and culture. In the 1980s, anthropologist Iégor Reznikoff noticed that the density of Palaeolithic cave art appeared to be proportional to the sonic quality of the site. Small red dots at certain sites seemed to be markers for maximum acoustic resonance, perhaps pointing at which sound-based rituals once served to merge body and earth: 'Because of the resonance, the whole body is implicated, sometimes in a subtle way. The approach is essentially physical; in this respect, we may say that the sounds and the whole situation are primitive. Since both the body and the cave vibrate we can speak of an earth or mineral meaning of sound.'

Ritual. Incantation. Song. Within chambers adorned with images of animals, art and sound may work together to conjure forth the more-than-human world. Reznikoff on the power of depiction and mimicry: 'taken by the power of these animal sounds and imitating them, you cannot help identifying with the corresponding animal... It is sometimes frightening, due to the power of the identification; this reminds us strongly of shamanic possessions or trances.' In the right cave. At the right spot. Clapping becomes hoof beats. Echoes blur and reverberate to conjure a herd of prey, stampeding. Torchlight and shadows flicker across the outlines of wild horses, elk, mammoths. Hunter and hunted commence to dance a dance of what is to come or in honour of what has always been, time and again.

Cross-modality information transfer is a term proposed by linguist Shigeru Miyagawa to describe the flow of knowledge between the auditory and visual centres within the growing brains of early humans. From the interplay of creation and enactment, a transmutation occurs. The abstract arises. Perhaps within the cave was where our species' symbolic minds began to take shape. Nestled within the blind wombs of the world, the first sparks of human imagination ignite.

'Songs of the Caves', a project headed by Rupert Till, aimed at exploring the acoustic fingerprints of five sites in the Altamira cave system in northern Spain. In addition to mapping each interior to assess Reznikoff's claims of a relationship between cave sounds and art, Till's team of musicians and archaeologists brought along period-authentic reconstructions of instruments to be played and recorded within the spaces. In the bison chamber of El Castillo, the low whirl of a bullroarer, a leaf-shaped stone that is swung on a string around one's head. In the deer chamber of Las Chimeneas, the thrums and slight echoes of a shaman's drum. Till notes in an interview that, at one point, someone began playing a replica of a 40,000-year-old vulture-bone flute in utter darkness. 'The music seemed to bring the environment to life,' he says simply. As it might have done the same coming from another flute, freshly carved and played for the first time in another cave yet to be discovered, long, long ago.

5

Reconstruction of the Paleosonic: Bringing Back the Sounds of Past Life

Museum of Modern Art, New York, USA | Yamal Peninsula, Yamalo-Nenets Autonomous Okrug, Russia | Wrangel Island, Chukotka Autonomous Okrug, Russia

It is a shame that sounds waves fade away, American writer Amy Leach muses in her essay 'Goats and Bygone Goats'. If they did not, 'the world, full of past sound, would be like the sky, full of past light. The world would be like the mind, for which there is no once.'

There is no discipline dedicated to reviving extinct sounds. While museums can create dioramas of past ecosystems and taxonomists can articulate bones of past creatures, vocal instrumentation – windpipes and lungs, tracts and palates – comprises soft tissue that leaves no trace in the fossil record. No matter how lively and vibrant, ancient soundscapes co-created and embodied are lost soon after utterance, dissipated into the winds of time.

But what if there was a way to bring back this most ephemeral of senses? So asked French artist Marguerite Humeau in her 2011 Museum of Modern Art exhibit titled 'Back, Herebelow Formidable'. After consulting with experts that included the surgeon who performed the second-ever successful larynx transplant, Humeau synthesized a new voice box for Lucy, the famous 3-million-year-old Ethiopian hominid. The contraption is made from SLA resin, acrylic, rubber and solenoid valves. The hiss in the background – air being run through a black motorized air compressor to simulate as breath. The wail that escalates – more human than one thinks possible from a Frankenstein machine so mechanical, almost abominable.

Humeau's *Opera of Prehistoric Creatures* also featured the resonance cavities and calls of a hell pig, a walking whale and an imperial mammoth. 'Mammoth Imperator' is 6 metres (19½ ft) long; its vocal cords are almost 20 centimetres (8 in.) wide. Exhibition-goers could feel the growl through their bodies and the bass through their chests. Children reputedly ran away in fear. Resources drawn to create this reconstruction: CT scans of modern-day Asian elephants from Berlin's Institute of Wildlife Research; insights from Steven Spielberg's adviser on the dinosaur calls in *Jurassic Park*; and exchanges with a French explorer credited with examining frozen woolly mammoths on Siberian expeditions, including Lyuba, a month-old calf that suffocated in the muds on the Yamal Peninsula 42,000 years ago.

Unlike Lucy, the deep, guttural rumbles emanating from Humeau's subwoofer-enhanced installation conjures a creature going about daily life. Only we, situated in its far future, interpret its sounds to be part of a dirge. Humeau's work seeks to merge scientific fact with imagination and speculation to evoke 'strong physical, almost supernatural experiences'. What scientists know: the last mammoths lived and died 4,000 years ago. While ziggurats were being raised in Sumer and the Pyramids of Giza were under construction, a small, isolated population lingered on in twilight on Wrangel Island in the Arctic Ocean, half a world away. What researchers suspected: the end came quickly for them, possibly due to a single extreme icing event that prevented the animals from accessing enough ground forage. Sparked by Humeau's work, we are now tasked to conjure new tales for one forever relegated to the past. What was said in the final call of the last mammoth, ringing across a realm of fir and whirling snow? Did it trumpet out in vain, in hope for an answer that would never come? And how should we ourselves deal with this process of crossing from presence into absence, towards that looming threshold which we must one day breach?

6

Calling Out Across Far Distances: Sound Messages in Air and Water

Pitt Rivers Museum, Oxford, UK | Bermuda Triangle, Atlantic Ocean

When a *Charonia* snail spots a bright red starfish, she will silently creep up to it, slithering over corals and rocks. Being slightly faster than her prey, there is no escape for the starfish. The snail will crawl on top of it, press it down hard with the muscles of her one foot, then stick out her radula, a serrated tongue, and start to saw into the meat. Having penetrated the flesh, she will release the paralysing poison in her saliva and start devouring at her leisure. First the delicious parts, the sexual organs and the gut, then on to more. The only escape for the starfish is to let go of one arm, hoping to cover some distance, before the snail gets hungry again. On a side note, she is doing good for her ecosystem: crown-of-thorns starfish, which can grow up to 80 centimetres (31½ in.) in diameter and sport venomous spikes, have no other natural predators. The snail is maintaining a healthy balance in her designated coral reef home by feeding on them.

Washed ashore, the dead husk of a *Charonia tritonis* is known as Triton's trumpet, or Triton's snail, her namesake no lesser than the son of Poseidon. On antique mosaics and tiles, Triton is depicted as blowing into this shell to calm the waves.

There is a strength and urgency to the conch shell's sound, echoing through its various lustrous nacre chambers, against its hollow calcium carbonate hull, spiralling up its cone-shaped structure that is said to reflect the motion of the planets. The conch shell is treasured among the religious leaders of island, sea and mountain cultures from Polynesia to Tibet, from Japan to Mexico. The iconic Hindu god Vishnu holds her in one of his hands and Quetzalcoatl, the Aztec god of wind, carries her around his neck. She is used as an instrument to call out to other worlds, to communicate with entities of nature, or to summon rain.

～～～

When the fir trees of the higher Alps in Switzerland grow from the rocky and steep slopes, they shoot out at a right angle to the hill, clinging with their roots into the soil. Only later do they lift their spirit and grow upwards towards the sky, the sun. What remains is a natural arch at the bottom of their trunk, a curve, that tells a story. Cut and hollowed, a tree trunk like this will make a long trumpet-like instrument, the iconic Alphorn. When lips blow into a mouthpiece at its top, the sound waves travel inside its tube to the flaring arch and out into the world.

The tunes emitted are in the natural scale, soothing and harmonic, possibly a bit simple for listening complexity but always pleasing. When from the mid-1900s it became fashionable for city people to reclaim a sense of lost nature in the Swiss mountains, the Alphorn got a bigger story than it might deserve. Lore spread of lonely herdsmen spending their summers alone on the alpine meadows and playing longing tunes for their beloved ones in the valleys below. These were heroes, it was claimed, who once blew loud and urgent from their lookouts, to warn of enemies approaching. This might all have happened once or twice, but most likely the horn was used to communicate with cows. 'They live in a different time-feel-zone and the low humming sounds of the horn seem to penetrate into their sphere,' says German artist Jonas Ried, who has built his own Alphorn to visit local cowsheds. 'They always react, unless they have stomach problems.'

Up to the present day, warning of imminent danger is still the responsibility of church bells in some places. The Austrian provincial law consolidated in the state of Vorarlberg declares: '§4 Bell ringing for storms: The following signal shall be used for warning and alerting with church bells: Storm ringing in the duration of at least 20 minutes.' There are some Austrian church bells that are directly connected to the Ministry of Defence and can be activated with a single click. In 2022 hackers gained access to the bells of St Stephan in Vienna and rang them at 2 a.m. for 30 minutes, until the priest switched them off manually – with his tablet.

If the message is elaborate and its destination far away, small notes can be tied to the legs of birds. Carrier pigeons are increasingly being used again as they can guarantee something that no Internet service can claim any more – secret delivery. It has been found that loud carrier pigeons reach their destination better than quiet ones. In China, the *geling* or *geshao* are elaborately carved whistles that are attached to the tails of carrier pigeons to protect them from attacks by birds of prey. The wind of flight is brushed through the whistle, sending a warning to other birds and, as a side effect, enlivening the human habitat with an Aeolian soundscape. The German sinologist Berthold Laufer wrote in 1908,

> [When] the birds fly the wind blowing through the whistles sets them vibrating, and thus produces an open-air concert, for the instruments in one and the same flock are all tuned differently. On a serene day in Peking, where these instruments are manufactured with great cleverness and ingenuity, it is possible to enjoy this aerial music while sitting in one's room.

The Pitt Rivers Museum in Oxford is home to 78 bird whistles from China and Indonesia. Inspired by them, former artist-in-residence Nathaniel Mann has designed new shapes and tonalities, enabling listeners to 'hear the flight of the birds described in sound'.

While such super-sounding pigeons can travel long distances to deliver messages, there are animals on our planet that communicate faster and over much greater expanses: Once upon a time, the ancestors of fin and blue whales lived on land. Adaptation to life in the water also meant the evolution of a new form of communication.

In the deep sea, at a depth of about 1 kilometre, lies the SOFAR or deep sound channel, created by the increasing salinity and decreasing temperature of the water. Humans first discovered it during the Second World War and spied on each other through its sound-conducting properties. Played at the right frequency, this sound channel is capable of transmitting sounds over a distance of 21,000 kilometres (13,000 mi.).

Whales know how to use this phenomenon for their own living environment. A male fin or blue whale singing into this sound channel off Ireland can hear its whale relatives off Bermuda 20 minutes later. A response would reach him after only 40 minutes.

'When whales use their songs to communicate with each other, they do so not only across space, but also across time,' says Chris Clark, a researcher at Cornell University's bioacoustics research programme who, with Roger Payne, discovered the whales' long-range calls. 'They repeat the same sound over and over again, and at very precise intervals . . . If you wanted to design a signal that could communicate across oceans, you'd come up with something similar to blue whale song.' It is believed that they are using it to create accurate mnemonic maps of the ocean floor.

The oceans, a world of sounding messages. Not only to listen to them, but to feel them with the whole body, to dive into a world full of vibrations – we are still far from a full understanding of what this means.

7

Reverberations Across Time and Fate: The Oracle at Dodona

The archaeological site of Dodona, Greece

October 2021. The theatre at the archaeological site of Dodona is currently undergoing restoration. Visitors are not allowed to walk up to the *summa cavea*, the top seating section once open to women and children. But they are encouraged to walk into the performance pit, to gaze up and envision the 55 rows of seats in their original splendour.

There is a spot in the orchestra area, dead centre, where one's voice projected out reverberates back into one's body, generating a resonance that seems to amplify one's conviction; or to shore up one's courage, as the condemned might have needed when pitted against animal combatants during the reign of Caesar Augustus in the first century AD, when the Romans converted the theatre into a gladiatorial arena for *Damnatio ad bestias*, death by wild animals. Perhaps such confidence can also be transferred through sounds in reverse, en masse and further back in time, channelled from the more than 17,000 spectators of the same theatre space, into the actors performing during the rule of the Greek king Pyrrhus in the third century BC.

Sound may not only bolster, but portend. But which ones will speak truth? Phyllomancy comes from the Greek words *phyllo* (leaf) and *manteia* (prophecy). Just east of the theatre in Dodona stand still older stone ruins of a sacred house, dedicated to the Oracle of Zeus. Unlike the more famous Oracle of Apollo at Delphi, Dodona priests arrived at answers not by inhaling earthly emanations, but by decoding the rustling of leaves from sacred oaks and the coos of wild doves. By the eighth century BC, a ring of bronze cauldrons on tripods was set around the tree. When struck, they resounded to form a barrier of protection against evil.

In legend and in life, those living throughout Magna Graecia consulted the oracle for answers. Achilles reputedly prayed for the safe return of Patroclus, his dearest comrade – a request denied, possibly because he consulted remotely. Odysseus wove a tale around asking the oracle whether he should return to Ithaca as himself or in disguise. Jason was told to place a branch from the blessed tree on the *Argo*'s prow as a direct divine connection on his quest for the Golden Fleece. More mundane queries from more mundane folks were carved on to lead tablets and have thus endured as long as epics: questions around trades and debts. Decisions around marriages and dowries. Worries about livestock and living. There once was a man named Hermon who asked which god he should pray to for Kretaia to bear him useful offspring. Answers arrived for large and small matters alike.

Back, back, further back. Before gladiatorial contests and stage plays, before the raising of stone houses and the cult of Zeus, pilgrims journeyed here to this valley on the eastern slopes of Mount Tomaros to seek advice from not only a single oak, but a sacred grove. Not much is known except that worshippers of the great goddess of fertility and abundance divined meaning from the ever-changing sounds within the grove, where trees low-whispered with each other alongside the breeze; and that the priests honoured this gift of foresight by sleeping on the ground, and by never washing their feet, so as to better connect with Mother Earth.

The silence of stones outlasts even the longest song of trees. There has not been a grove at Dodona for many an age. Cities and empires rise and fall like waves, doomed to repeat and fade away. Inscribed questions have been answered and recovered, shipped off to Athens as artefacts of antiquity for exhibition. Yet an oak still stands today at the archaeological site of Dodona. Perhaps this descendent is a mere remnant and reminder of its ancestors' glory. But maybe the power of an oracle is immanent in all oaks, and that even now, the tree is translating the gentle rains that fall upon this valley on the eastern slopes of Mount Tomaros and is willing to offer counsel to a new generation, if we would only learn to listen anew, recall the pattern.

8

Word on the Wind: The Curious History of the Aeolian Harp

Hohenbaden Castle, Baden-Baden, Germany | The Aeolian Islands, Sicily | Golden Gate Bridge, San Francisco, USA

'The Aeolian harp is a musical instrument which is set in action by the wind. The instrument, which is not very well known, is yet very curious, and at the request of some of our readers we shall herewith give a description of it.' This description is from an entry found in the 483rd issue of *Scientific American Supplement*, published on 4 April 1885 – sandwiched between sections on 'Improvised Toys' and the 'Manufacture of Illuminating Gas'. Within the short piece is an attribution of the first box-shaped model mounted with catgut strings to a German Jesuit savant named Kircher, dated to 1558; sketches and descriptions of subsequent refinements, including an especially famous version at Hohenbaden Castle in Baden-Baden; an addendum: instruments that achieved spontaneous resonance when exposed to air currents most likely struck 'observers of nature in times of remotest antiquity'.

Indeed, mentions of songs through strings played by no mortal hand reach further back in history and myth. Musicologist and instrument collector Carl Engel speculated that the lyre suspended over the pillow of King David in the Bible was likely a Hebrew kinnor, a portable instrument that woke the king at midnight when the north winds blew. The term 'Aeolian' itself was derived from Aeolus, Greek king of the isle of Aeolia. Master of the winds, he offered a homesick Odysseus the means back to Ithaca in the form of a great boon: an ox-hide sack sewn with silver thread with the power to restrain all winds except the one blowing west – towards home. Yet the greed in men's hearts proved to be their undoing; just as the shores of Ithaca came into view, Odysseus' crew opened the bag, believing it to be filled with riches. With the trapped winds unleashed, their ships were blown back by the storm to whence they came. Odysseus would not set foot on his home shores for many more years. Perhaps here the term 'aeolian' serves as both description and cautionary tale: not by the ambitions of men should the will of winds be sullied, unless one wishes not for a smooth voyage, but a tragic one.

The instrument that should not be played; lost and then rediscovered, popularized again. Tinged with a touch of the divine, the Aeolian harp became a touchstone for the Romantics. Back in *Supplement* No. 483, an account of French composer Hector Berlioz, describing the melancholy of an Aeolian harp played in the dead of winter, defying the reader not to experience 'a profound feeling of sadness and of abandon, and a vague and infinite desire for another existence'. English poet Samuel Taylor Coleridge spent three decades working his blank verse 'Eolian Harp', weaving it into a meditation spanning from love and marriage to his wife Sara to an exploration of whether living things can be considered organic instruments, framed as various forms but all brought into being through a single tremble of holy thought. In 'Maiden Song of the Aeolian Harp', American Transcendentalist Ralph Waldo Emerson speaks on behalf of the instrument sitting by the western window of his study: 'Keep your lips or fingertips/ For flute or spinet's dancing chips/ I await a tenderer touch/ I ask more or not so much: Give me to the atmosphere.'

Atmospheric music is aleatoric music – compositions left solely to whim and chance. Yet unlike the percussive chaos of wind chimes, the wind harp contains an underlying order. When a hand plucks a string, the sound created is mainly a fundamental note accompanied by harmonic overtones. When the wind plays a string, only the overtones reverberate. Thus the ghostly, ethereal pitch.

Not only does the player need not be present, sometimes even the instrument itself is immaterial. Emerson's protégé Henry David Thoreau found otherworldly thrums neither in instruments nor the natural world he sought refuge in, but rather in the latest technological scourge to intrude upon his deliberate nineteenth-century existence at Walden Pond. A journal entry on 3 September 1851: 'As I went under the new telegraph wire, I heard it vibrating like a harp high overhead.

It was as the sound of a far-off glorious life, a supernal life, which came down to us and vibrated in the lattice-work of this life of ours.'

In our modern-day, faster-paced world, one hub for Aeolian sounds seems centred around San Francisco. At the city's Exploratorium stands an 8-metre-tall (26 ft) Aeolian harp with seven stretched strings. Sculpture artist Doug Hollis designed it to take advantage of the natural wind tunnel between Piers 15 and 17, which seems to pick up in tone every mid-afternoon. And on a breezy day in South San Francisco, the wind rings through the arched and rusted beams of a 28-metre-high (92 ft) sound sculpture designed by Lucia and Aristides Demetrios.

Yet these monuments to the wind's words pale in comparison to another in the Bay Area. A 2020 retrofit of the sidewalk safety railings on the Golden Gate Bridge's western side made it safer and more aerodynamic for withstanding high winds, but had the side effect of turning the landmark into a giant singing harp. Some locals find the tone peaceful and meditative. Others living miles away feel they might be driven mad. Music is inherently subjective, and even an inadvertent performance absent any performers will have its devotees and detractors.

9

To Be Defined By Sound: Regarding the Cicada

'Ulahi and Eyo: bo Sing with Afternoon Cicadas', Papua New Guinea | 'Cicada in Malaysian Rainforest', central Malaysia | 'Moccasin Game Song: Cicada or Locust Song', the Navajo Nations, New Mexico and Arizona, USA | 'Call of the Cicada' (Utom udei lumen helef), southwestern Mindanao, Philippines

Play a game of association with insect sounds and note what arises. The thrum of bees – perhaps the image of wild flowers, teamwork, the promise of honey. A mosquito's whine – maybe the tensing of muscles, a heightened paranoia. Chirping crickets can compel one to bound forth into the wide yonder, across grassy fields, into long hot nights. Rarely do we dwell long on the insect sound itself, save perhaps those coming from the cicada. They do not sting, do not bite, do not leap away in startlement. Song is the only means they have to assert themselves upon the world, and by extension upon us, via their sheer volume and vitality.

Throughout history humans have been entranced by these bugs, winged and red-eyed, emerging from the ground en masse. The first-century BC Greek poet Meleager of Gadara dedicated an ode to the cicada as 'the shrill-voiced and dewdrop-sweet'. The twentieth-century Chilean poet Pablo Neruda offered his own tribute to another sending 'its sawing song/ high into the empty air'.

To many, the connection between the cicada's song and its impending doom is inextricable. In Plato's *Phaedrus*, Socrates spoke of the origin of cicadas as former humans who lived before the Muses and became so enamoured with the advent of song that they forgot to eat or drink and thus perished without realizing it. The fifth stanza of Raymundo Pérez y Soto's 'La Cigarra' ('The Cicada'), a popular Mexican song in the Mariachi tradition, reads, translated: 'Under the shade of a tree/ and to the beat of my guitar/ I sing this huapango happily/ because life ends/ and I want to die singing/ as the cicada dies.'

Desire so great as to spill forth and overwhelm life itself – this is perhaps part of the romance and allure. Such a long age to spend below ground, alone, in darkness, sucking the sap of roots, preparing for those brief days of song: two to five years for annual varieties; up to seventeen for periodical broods. Perhaps to emerge in full glory into a world so utterly different from whence one once came exemplifies courage. Perhaps one must cling to some universal truths: the need to serenade another. The need to come together.

For male cicadas, no mere rubbing of parts together suffices to generate their sounds. Twin tymbal organs, specialized drums with ribs that flex and click; a hollow abdomen, a resonance chamber to amplify; wings for directing to those who will be listening. The body is itself a musical instrument. By himself, the male cicada is a lone star, as loud and insistent as a jackhammer. Together, in chorus with thousands, millions, trillions of his brothers, his rhythms turn into a force of nature. Landscapes and ecosystems are transformed via his symphonic and bodily contributions. Here is to becoming sustenance, ambience, vitality. Listen. Here is the distilled essence of summer.

Smithsonian Folkways Recordings, the non-profit record label of the Smithsonian Institution, features folk music inspired by cicadas from cultures across Southeast Asia, Oceania and the Americas. Titles featured include: 'Ulahi and Eyo: bo Sing with Afternoon Cicadas', from the first generation of Bosavi people in an independent Papua New Guinea; 'Cicada in Malaysian Rainforest', from the spiritual dream songs of the Temiar people of the central Malaysian rainforest; 'Moccasin Game Song: Cicada or Locust Song', from the Navajo Nations of Arizona and New Mexico; and 'Call of the Cicada', from the T'boli, an Indigenous group living in scattered villages across southwestern Mindanao, Philippines. This last song, played on a bamboo string instrument called a zither, alludes to the 'unrelenting and shrill calling of the cicada at sunset'.

A source of spiritual inspiration for some, an intolerable cacophony for others, and everything in between. What is undeniable is that the way of the cicada is one of fierce exaltation against the drudgeries of an earthbound existence.

Perhaps this is what resonates, soars long after the notes and the insects have perished. To give wholly one's being to what is worthy of love – is that not what we should seek to strive for in this or any life?

10

Gagaku:
A Music that Crosses Space and Time

Tokyo Concert Hall, Japan | Sazare-ishi in Hyuga City, Japan

'I feel it would be ideal if my music could sound, and then when the echoes of these sounds come back, I would no longer be there,' mused Japanese composer Tōru Takemitsu. While we think of music as something transient that activates the space around and within us by a narrative explored in rhythm and melody, Takemitsu's ideal music would carry only an 'immeasurable metaphysical sense of time'. A music that allows us to forget about the boundaries of our world. He found this quality in Japanese *Gagaku*.

Gagaku translates as 'elegant music'. Played at the Japanese court in a stylized manner by musicians in opulent silk robes, it is still performed to this day. The *Gagakuryo* or Imperial Household Ensemble was founded in the year 701. The songs are handed down from family member to family member, having arrived currently with the 36th generation.

To learn to play *Gagaku*, one first needs to discover one's chosen instrument's character. For a year the student taps its rhythm on their knees while evoking each instrument by a different vocabulary. This unique practice allows them to understand the timbre, the spacing between notes, rhythm and melody before even touching the instrument itself. Eventually the secret parts of the music will be handed down, from teacher to student, not by words, but by intention.

'There is freedom for me within the strict confines of the already set music of *Gagaku*,' says Ko Ishikawa, a Japanese *Gagaku* player who was into British punk and noise music before he got spirited away by these ancient sounds.

Once, *Gagaku* was a part of the Tang Empire, where it had arrived as echoes from the arid landscapes of the kingdom of Khotan (modern-day Iran), the lush forests of Vietnam, the sandstone hills of India and the rugged mountains of the Korean peninsula. What was lost and what was added on *Gagaku*'s long journey through space and time? Ko Ishikawa elaborates:

> *Gagaku* is a big tree over a thousand years old. Every year this tree has a splendid blooming and some branches are dying. I am only a small fragment of consciousness reflecting the appearance of this tree; my subjective impression is not important. I just let the flowers appear as they blossom.

Gagaku

The *Tama-shizume* is a concept in Shintoism that translates loosely to 'the repose of the soul'. The idea is that the soul wearies over time. But *Gagaku* can revive it.

The *Gagaku* ensemble's most exhilarating instrument might be the *sho*, both in looks and sound: the musician inhales and exhales through a mouthpiece into seventeen bamboo pipes, oscillating the air into a sonic vortex, thus creating a never-ending tone that, according to Takemitsu, makes it 'possible to imagine the concept of transitoriness but not necessarily that of lifelessness'. The *shō* was mentioned in the *Taigenshō*, a book on music from the Japanese Middle Ages: 'In the midst of the dazzling autumn flowers, an ethereal man, with formal trousers, gently holds his sword and raises his hand to blow.' A continuous thread of breath connects autumn and spring, this world and the other, the *Kakuri Yo*, the deep, dim realm of spirits and gods. Distances of temporalities and spaces are crossed by sound waves, weightless and invisible. Some say the *shō*'s sound resembles the songs of the mythical phoenix.

The flutes and percussion instruments used in *Gagaku* hint at a rhythm that disappears as soon as one tries to hold on to it. The string instruments with their gentle wind-like tones give accentuation. *Bugaku*, a dance, might accompany the orchestra, with choreographies so slow that the dancers seem to have frozen in time. Sometimes there is also *Kagura-uta*, a song, intense and haunting, like a spell cast, unbroken.

In football stadiums around the world the Japanese sing their national anthem, translated: 'May your reign continue for a thousand, eight thousand generations, until the tiny pebbles grow into massive boulders lush with moss.' The tune was partly composed by *Gagaku* players from the imperial court. The words are based on a tenth-century poem describing a *sazare-ishi*, a rock formation where small pebbles grow together into a boulder, before lichen and moss will transform them back again into soil.

In Japanese aesthetics, everything arises from and dissolves into nothingness. Nothingness is not an empty space, but a space full of potential. *Gagaku* seems to dip into this sea. For what is music but a presence in which we immerse ourselves? A wave rolling by, fading, dispersing, then building up again. Not to be held, but to be enacted.

> *After I got off the train and was walking toward my home, the new green of the hedgerows was sparkling, and at the shrine along the way, misty vapour seemed to be rising into the sky from its large tree. At that moment it was not only the* shō *that the sound resembled but something closer to the image of the entire* Gagaku *ensemble. I felt the sound of the large taiko drum seemed to reach into the ground like roots and give order to the earth. The metallic* chi-chi *sound of the small shoko gong sounded like it was transmitting to the other world and the* took-toko *sound of the kakko drum sounded like the beating of time passing. The sound of the ryuteki flute was like a flow of air that connected the trees to the sky, carrying the wind with it. And the hichiriki flute was like the voices of us living creatures on the earth, like the voices of human beings*

and the voices of animals, and the sound of the shō' *pipes had a resonance of sunlight from the sky reflecting off the myriad things of the natural world.*

MAYUMI MIYATA

11

Sounds of the Cryosphere

Polar regions | Cryosphere

Dove blue, grey like pearls or smoke, volcanic black, granite, some like raw jade. Depending on the thickness of the ice, its proximity to the water: aquamarine, milky blue shading into full navy blue. Young fractures: glittering green-blue. Old blunted icebergs: more grey. At sunset: icebergs catch the light in pink, reddish-yellow, in watered-down purple and soft pink.

Writer Barry Lopez sings the landscape of the polar regions alive in his 1986 travelogue *Arctic Dreams*. Sound here, however, rises only from the animated world, the 'tremolo moans of bearded seals. The electric crackling of shrimp. The baritone boom of walrus. The high-pitched bark and yelp of ringed seals.' What then is the sound of ice?

Friday, December 20th, 1850: Nature is suspended and silence reigns, so that one can hear the clock in one's pocket. Meanwhile, the celestial phenomena offer many interesting things: the stars shine in the middle of the day, ring-shaped rainbows are often seen around the moon, shooting stars and meteors are not rare.

Missionary Johann August Miertsching described an awe-inspiring world, but his Arctic landscape once again stayed silent in the minds of readers; many fevered for the first-hand reports published in magazines and periodicals, describing a sublime environment that befit the Gothic longings of the times. While the first and very few expeditions starting in the fifteenth century ventured out from the White Sea, it was only in the nineteenth century that ships like Miertsching's HMS *Investigator* sailed out further to measure the sea beyond. Their crews were augmented by passengers skilled in botany, astronomy, hydrography, mineralogy, statistics, zoology, meteorology and the visual arts, but with no tools or knowledge to capture sound.

If the resonances of the ice should be told, someone had to find a vocabulary for it. There is a line in the Exeter Book, written in England in AD 980, that plays onomatopoeically with the sonority of frozen water: *glisnaþ glæshlutter gimmum gelicust* are words that don't have to be understood. It's enough to pronounce them to elicit a sense of their meaning: 'I Is byþ oferceald, ungemetum slidor,/ glisnaþ glæshluttur gimmum gelicust,/ flor forste geþoruht, fæger ansyne.' (Ice is very cold and immeasurably slippery/ it glistens as clear as glass and most like to gems/ it is a floor wrought by the frost, fair to look upon.)

Squinting may help to hear the crackling and bursting that ooze out of Caspar David Friedrich's masterpiece *The Sea of Ice*, offering a synaesthetic experience; the painting provides a glimpse into a different reality to that portrayed by those who came back from the polar regions to tell of what they saw, and is based on first-hand reports scribbled down in notebooks that survived their authors. Ice turns into a cold and indifferent texture and the shields of its glistening beauty fill in the gaping voids, cracks and crevices of an emerging sonic imagination.

It was the discovery of polar explorer George De Long's stiff, frostbitten hand, sticking out of the snow in Siberia's Lena Delta, that revealed his diary, later published by his widow. 'The pressure was tremendous and the noise was not calculated to calm one's mind. I know of no sound on shore that can be compared to it.' For two years he had been in the ice drifts. When there were sounds, danger was imminent. 'And however beautiful it may be from an aesthetic point of view, I wish with all my heart that we were out of it.' It was his descriptions that later inspired explorer Fridtjof Nansen to build ships that could withstand the pressure of the ice and make use of the natural forces of the transpolar drift. He, in turn, didn't mind the horrid noises: 'I laugh at the power of ice. We live in an impenetrable castle.'

Today it is the icebergs that break apart, not the boats travelling there. Accelerated by rising temperatures, so-called icequakes have become common. The death throes of a 59.5-kilometre-long (37 mi.) iceberg recorded near the tip of

Argentina 'emitted what sounded like tortured groans, shudders, and cracks as it ran aground, spun around, scraped along the seafloor, and then broke up over open water.'

While the ice is melting, scientists, musicians and poets rush to capture its fading echoes, finding melodies along with narratives. With the help of hydrophones they are astonished to hear 'bubbles making a bloop, tick or pop sound . . . that lasts 10 milliseconds or less'. Released from the melting ice, these bubbles fill the soundscape of fjords, attracting seals, which hunt better in the noise. When the glaciers have completely retreated to land, the seals leave, having no place to hide from the sonar of killer whales.

Some of the bubbles released from the ice are thousands of years old. Compressed chambers of air from a different time and age, locked into the cold. When they burst, their memories disappear.

Our memories will not be trapped in ice. Of that, the future is almost certain.

12

The Stories and Secrets of Singing Sands

The Lop Desert, Xinjiang Uyghur Autonomous Region, China | Copiapó, Chile | South Sinai, Egypt | Dunhuang, China

1271. A seventeen-year-old Marco Polo sets out with his father Niccolò and his uncle Maffeo to attend the court of Kublai Khan. Later in life the Venetian would dictate the details of this life-changing journey to romance writer and fellow Genoese prison inmate Rustichello da Pisa, conjuring into existence that most enduring of chronicles: *Il Milione*, more commonly known as *The Travels of Marco Polo*.

Chapter xxxv: Polo describes crossing the deserts of Xinjiang and the voices reputed to emanate from those sands that lure travellers astray. From 'Of the Town of Lop – Of the Desert in Its Vicinity – And of the Strange Noises Heard By Those Who Pass Over the Latter', an excerpt: 'Marvellous indeed and almost passing belief are the stories related of these spirits of the desert, which are said at times to fill the air with the sounds of all kinds of musical instruments, and also of drums and the clash of arms.' A mere folk-tale perhaps, but one that matches the tone of the rest of the travelogue: alluring, fantastical, yet always grounded in some seed of truth, beheld by the eyes of a soul ever observant, ever curious.

~~~

1835. A 26-year-old Charles Darwin has been conducting surveys in South America for more than three years. After a gruelling expedition in the Andes, he returns to HMS *Beagle*, stopping at the northern Chilean town of Copiapó.

The local talk and obsession around mining proves numbing; his interest is instead piqued by accounts of 'El Bramador', translated as 'the roarer or bellower', a local sandy hill that would boom whenever people climbed it.

From Chapter XVI of *The Voyage of the Beagle*: 'One person with whom I conversed had himself heard the noise: he described it as very surprising; and he distinctly stated that, although he could not understand how it was caused, yet it was necessary to set the sand rolling down the acclivity.'

Connection and synthesis: Darwin would wield his gift for both a few months later on the Galapagos, cementing his legacy in the annals of biology. For now, the young geologist recalls a similar sound on a low sandstone hill in South Sinai, as described by explorers Ulrich Jasper Seetzen and Christian Gottfried Ehrenberg. Seetzen noted the tone as first taking on the strains of an Aeolian harp, before becoming that of a hollow top, lastly rising so loud that the whole earth seemed to quake. Ehrenberg confirms this on a subsequent visit, 'beginning with a soft rustling, it passed gradually into a murmuring, then into a humming noise, and at length into a threatening of such violence that it could only be compared with a distant cannonade, had it been more continued and uniform.'

China. Chile. Egypt. Corroboration at multiple locations renders Polo's singular account of desert hauntings five centuries prior less fanciful. But what lies at the heart of this singing and humming, booming and roaring? The shifting sands do not reveal their secrets easily.

～

2007. Physicist Simon Dagois-Bohy and his team seek answers by sliding feet-first down dunes and triggering sand avalanches near Tarfaya, Morocco, and then a year later in Al-Ashkharah, Oman. The key to sandsong, it seems, lies within each grain, tumbling and colliding in concert with others to create a synchronized sound. While the dune face serves to amplify the volume, it is not vital; researchers were able to replicate the pitch with samples brought back to their lab in France.

Some essential conditions: the sand grains must be dry and round, must contain silica, must range between 0.1 to 0.5 millimetres in diameter. Uniformity in size affects singing quality. The homogenous Moroccan sands wailed pure at a steady 105 Hz – in the neighbourhood of a G sharp – while the Omani sample, with widely varying grain sizes, wailed across nine notes from which a single frequency could not be discerned.

Only thirty to forty sites around the world can be coaxed into song (though many beaches have been heard to squeak). Among them, the 3-kilometre-long (2 mi.) dune in Kazakhstan's Altyn-Emel National Park; the 200-metre-tall (656 ft) Eureka Dunes in California's Death Valley National Park. Along the Silk Road east of Polo's Lop Desert and near the city of Dunhuang lies Ming Sha Shan, translated as Echoing Sand Mountain, now a popular attraction. After paying the admission fee and seeing the newly built 'Tang-dynasty' architecture and the Crescent Moon Lake, there is nothing left to do but to head out to the dunes. Options include travel by camel, by glider, by ATV or just walking. Shoe covers for avoiding the mixture of sand and camel dung are available for a small fee. The best spot for overnight campers to catch the sunrise is at the top of the western hill; the best time, before 6 a.m. If the wind blows true and you find yourself beyond the reach of summer crowds and the range of on-site speakers blaring music, there is a chance you can hear how hills and air work in concert to sound, to stir, to speak what they need to say.

13

The Humility Pipe

Bach's favourite organs in Eisenach and Leipzig, Germany | Vox Balaenae at the Beijing National Centre for the Performing Arts, China | The sound of the ancient Greek water-organ at the Acropolis Museum, Athens, Greece

Music and mathematics come together in Bach's *Well-Tempered Clavier*, a collection of fugues and preludes for young and experienced learners of pianos, harpsichords and organs. While its title refers to the logic, structure and play of harmonic intervals, some keyboard instruments might indeed be called fickle, moody and ill-tempered companions, if not properly taken care of. The sound of a harpsichord easily becomes fluttered and flat if the performance space does not suit it. Humidity and sudden changes of temperature bend its wood in unforeseen ways. Breathing concertgoers: Horror! A door kept open to let them in: Tragedy! Its musician will need to tune its keys with light, tender fingers before even thinking about starting to play. Do harpsichords even like having an audience, one wonders?

Pianos are built with a sturdier, heavier frame at the expense of portability. Ask a moving company for a quote for carrying your piano to your next concert space. And pipe organs? Once installed, they stay put. Raging wars or cold winters have no impact within the centuries-old stone walls the instrument resides in. Instead they permeate the building they were decreed to and rule it, filling air and space and every soul within with sound that can leap from transient to tragic within a beat.

While the origins of the *órganon*, Greek for 'instrument, implement, tool', go back to the third century BC, the organ seems to be the epitome of a German instrument, for good or for ill. With 50,000 organs in a country with 10,790 towns and villages and 170 practising builders, the craft of organ-making has been elevated by UNESCO as part of Germany's intangible cultural heritage. 'The organ is an expression of the landscape in which people live. A good organ is a mirror of their souls. Its sound makes people richer,' states Klais Orgelbau, the leading organ builder in the land of organ builders.

A pipe organ doesn't need an orchestra to create the sound of an orchestra. A single player masters the wind flow through the pipes, with each emitting a different pitch and timbre. Small pipes the size of a fingernail will sound like a high-pitched flute, while long pipes send out a spine-shivering bass. The acoustic energy brings the hollow spaces within the body to a tremble: deep tissues, beer bellies and double chins quiver in delight, awe or fear. And while classic pipe organs imitate classic instruments, pipe organs used in silent theatres can mimic the sound of a telephone, even as their sisters in department stores recreate the merry-go-round sounds of capitalism.

The organ player sits not unlike the navigator of a spaceship as they journey across realms of sound. Turning their back on the wide open space of the building, ignoring those below, unseen and unheard: the church congregation, the consumers, the movie-goers. The player wields their entire body: hands on the manual, feet at the pedals. A single person in singular control, steering airs and moods.

> I went below to the lounge, from which some chords were wafting. Captain Nemo was there, leaning over the organ, deep in a musical trance... The captain's fingers then ran over the instrument's keyboard, and I noticed that he touched only its black keys... Soon he had forgotten my presence and was lost in a reverie that I no longer tried to dispel.

Professor Aronnax's account of the organ playing at the bottom of the sea in Jules Verne's classic *Twenty Thousand Leagues Under the Sea* is not the only organ-playing villain in literature and not even the only organ-playing villain at the bottom of the sea. Davy Jones in the *Pirates of the Caribbean* film series is able to rest his troubled mind at the console. The organ players of movie-lore – Dr Jekyll and Mr Hyde, the

The Humility Pipe

phantom of the opera, Inspector Dreyfuss in *The Pink Panther*, or the *Carnival of Souls* – all unthinkable without the prodigy who has crossed the thin line that separates genius from lunacy. Is the organ amplifying the madness within the player or is the player becoming mad in attempting to control an instrument as powerful as the organ?

'I want more gravitas!' Bach would lament over most organs of his lifetime. In the eighteenth century only a few churches in Germany featured 9.75-metre (32 ft) organ pipes that were able to elicit *Erdenschwere*, a sound he liked, a sound that conveyed the heavy weight of the earth. At 16 Hz and at the boundary of hearing, such long pipes induced a feeling beyond musical scale.

Building anything bigger was impossible during Bach's lifetime; the pipe would collapse under the sheer weight of its tin and lead alloys. However, an apt player could combine the sounds of a 9.75-metre pipe with a 4.87-metre (16 ft) pipe in a clever way and arrive at a similar effect of a 19.5-metre (64 ft) pipe. Rumours hold that this acoustic trick was used during the masses, tasking the organist to produce a sensation that would give rise to an unheard, yet awe-inspired feeling in the body, a closeness to God for those within its acoustic realm: the so-called humility pipe.

Hermann Hesse picked up the German fascination with organs in his strange 1943 novel *The Glass Bead Game*, and one wonders what would happen if such an instrument ended up in the wrong hands: 'And this organ has attained an almost unimaginable perfection; its manuals and pedals range over the entire intellectual cosmos; its stops are almost beyond number. Theoretically this instrument is capable of reproducing... the entire intellectual content of the universe.'

Eventually, the twentieth century made it possible to equip organs with 19.5-metre (64 ft) pipes. One is now built into the Midmer-Losh organ at the Atlantic City Convention Center (with 33,114 different pipes, its builders aimed for an organ so complex that it cannot be fully played), or in the Pogson organ in Sydney Town Hall, then another at the Beijing National Centre for the Performing Arts: here the Vox Balaenae roars deep and mighty with a whale's voice. For those not able or willing to travel so far for an experience, an ill-tempered washing machine in high spin can create a similar infrasound. In the background, perhaps someone mad with household chores filling it with ever more clothes, secretly ruling the house.

14

The Untameable Ritual of Keening

West Ireland

Oh my beloved, steadfast and true!/ The day I first saw you/ by the market's thatched roof,/ how my eye took a shine to you,/ how my heart took delight in you,/ I fled my companions with you,/ to soar far from home.

Four hundred more lines follow these ones, evoking a man who no longer is and never again will be. Line by line, everything he leaves behind is put into words, wails and sighs: the memories that still linger in the air, the feelings he stirred, the relationships he forged, the anger and bereavement of those left behind. Loss. Void. His absence.

Passed on for decades in oral tradition, *Caoineadh Airt Uí Laoghaire* (The Lament for Art O'Leary) is one of the most haunting poems from the Irish isles, conjured up by a woman during the wake of her husband after he was murdered in 1773.

It was then sung at funerals and wakes, accompanied by high-pitched wailing, grieving, crying and growling – the deepest of sadnesses. And eventually became one of the best-known poems in England and Ireland.

Eibhlín Dubh Ní Chonaill grew up in County Cork, where the tides of the sea cast a constant echo on those who live nearby. Married off at fourteen she would return home only months later, when her much older husband died. One day, she meets the eye of Art Ó Laoghaire. They fall in love. Soon kids are born; they share a life. When his mare gallops back to Eibhlín alone she knows what happened. Led by the horse she arrives at the scene of his murder, scoops up his blood with her hands and drinks. Later, at the wake, the words will run out of her, as if they had always been in her, ready to be shared.

Keening laments traditionally were expressed in *rosc*, an ancient metre that has only survived in Ireland and Scotland. Eibhlín's song, beautiful and fleeting, not caught in writing, was passed on by professional keeners, *mná caointe*, women who would mourn and wail for a small salary during wakes and funeral rites. They served as vessels to keep an echo alive. The texture and weaving of a life once vibrant but now cold as stone. One woman's words pouring into another's body. Verse and rhythm passing through their flesh, tying places and times, from one to the other, one to the other, evoked and triggered by wails, screeches, raptures. It was a rite of passage, a guiding of the deceased from this world to the next. And it was also much more than that.

'Wailing is loud. The keening was a primeval scream, a calling out of the agony of death, an eruption of despair, tenderness, fear, love, loss and pain,' writes Kevin Toolis in his book *My Father's Wake*. One is reminded of the other rite of passage only women can provide for, that also comes with a howling and a screaming, untamed. One has to let go to let new things happen. The vocalization guides the body. Together with breathing it helps to find the portals. While we fulfil the most humane and sacred of acts, our voices will roar like animals. A ghost in the throat, as Irish poet Doireann Ní Ghríofa coins it.

One of the elements of keening is the chanting of old vocals, the *ochón*. The evocation of the hollow 'o' starts deep down in the belly, just below the navel. The singing of the *ochón* is said to lift the mind to a certain pitch, releasing the grief. Then other pitches follow, going deeper into the cellular structures of the body. To outsiders it might have sounded crazy but for those at the wake it was respect. Providing the deceased a safe passage to the other world. Only animals were buried without a proper wailing.

Keening as part of a funeral rite has almost disappeared in Ireland. Some say it started during the years of the Great Famine, when so many died and those left starving and suffering had no strength to honour each loss. Some emigrated to America, where the keening women and customs did not follow. While some of the keening laments like 'The Lament for Art O'Leary' were eventually written down and passed on in their own literary right, no Irish keening ceremony has ever been recorded and stored on tape. Their magic remains one that has to be experienced, the howling and screeching relieving the grief deep inside, letting it go, allowing the pain that comes with the loss to be voiced. And only then can new strength be found, with other songs, like *An tAmhrán Geal* (The Song of Light), an ancient Irish evocation with a tradition that reaches just as far back in time: 'I lie with the King of Creation/ May he lie with me./ His two hands beneath my head/ As I lie and rise again/ As I lie and rise again.'

15

Death of a Sound and Age: Pining for the Foghorn

Chebucto Head Lighthouse, Nova Scotia, Canada | Partridge Island Lighthouse, Bay of Fundy, Canada | Souter Lighthouse, Whitburn, England

February 2021. The final blast sounds from the foghorn at the Chebucto Head Lighthouse in Nova Scotia, Canada. For 120 years it has howled into the North Atlantic, warning sailors against ruin in conditions where sight would be of no avail. One last three-second wail, low and sonorous, before an eternity of silence.

Two hundred kilometres (61 mi.) northwest in the neighbouring province of New Brunswick is the birthplace of the first 'automated steam-powered whistle'. Walking home one foggy night in 1853, Glasgow-born Robert Foulis was guided by the tunes of his daughter playing the piano. But only the lowest notes reached his ear through the dense cloak of maritime mist. The inventor, artist and civil engineer would spend the next six years designing and advocating for a device that generated low-frequency tones to warn ships in low-visibility conditions. Safer and cheaper than constant cannon-fire, farther-reaching than clockwork bell-strikers, the coal-fuelled foghorn at the Partridge Island Lighthouse began operation in 1859. Yet it was a T. T. Vernon Smith who pitched Foulis's designs to the local commission. Foulis would receive no compensation for his invention, even after the House of Assembly in 1864 recognized him as the true inventor of a life-saving device which 'has been of great practical utility in the Bay of Fundy'. He died two years later in his adopted country, in poverty, buried without a stone to mark his resting place. His legacy would come in the form of a plaque placed posthumously on Partridge Island in recognition of his achievement, and in the foghorn which would sound for another 132 years, switching off in 1998.

'It is internationally recognized that sound signals like foghorns are not a reliable source of navigational information,' states Canadian Coast Guard Regional Director Harvey Vardy. In the twenty-first century, GPS systems allow boats large and small to find their way through that most disorienting of sailing conditions. Forty-six foghorns have been decommissioned across Atlantic Canada over the past decade. There is no longer a need to keep the nightly vigil, no financial case to keep the signal going.

Nostalgia then, which penetrates deeper into the mists of the soul than reason. For many, the link between life on the coast and the cries of the foghorn has become inextricable; few man-made sounds are capable of stirring so complex a suite of emotion. For the forlorn mariner, the sharp reminder of danger marries with the bright relief of homecoming. For the seaside community, associations of safety and reprieve, woven over the decades into rich tradition. Such a sound, beginning brutish and defiant up close, softened and rendered beautiful by distance and the unique contours of each landscape. Haunting, insistent and clarifying, the foghorn served as their hymn to the sea, that unfathomable realm they have long beseeched for bounty and mercy.

Like the fog it seeks to pierce, the cry of the foghorn does not endure. Where once was sound has now been stilled, a wave that has reached the shore, a journey completed. How to mourn something so wedded to the borders between land and sea? A mass, perhaps, to honour and remember the dead.

June 2013. More than fifty cruise ships and patrol vessels, personal yachts and lifeboats gather in the North Sea around the Souter Lighthouse in Whitburn, South Tyneside, England, syncing their horns to a brass band score and foghorn tunes performed live to an audience on shore. Orchestrated by artists Lise Autogena and Joshua Portway in collaboration with composer Orlando Gough, *Foghorn Requiem* sought to 'mark the disappearance of the sound of the foghorn from the UK's coastal landscape', to both celebrate and mourn an iconic sound that has come to embody local history and folklore.

Some in attendance, like music journalist Jennifer Lucy Allan, find themselves changed. Rocked by what she calls an 'aural obliteration' by being next to the foghorn's diesel-powered roars, Allan becomes obsessed with this sentinel from a bygone age, embarking on a quest to tease apart its deafening, wistful power. 'Nothing speaks to the sea like a foghorn does,' she writes in her book *The Foghorn's Lament*. 'Nothing gives such comfort when it warns, and nothing else comes imbued with the colossal weight of life and death, memory and melancholy.'

There is something resonant about a monument which once guarded so many against death, now confronted with its own mortality. Perhaps the mind cannot help but make the leap, to realize that we too shall one day fall mute, becoming husks littering a landscape of former purpose and glory. How then we might also yearn to bellow out at the world, to speak the sum total of our trials and triumphs through one sound and, in so doing, come alive again, one last time.

16

A Voice, an Echo, Silence

Echo stone, Brandenburg, Germany | Gol Gumbaz mausoleum, Bijapur, India | El Castillo Temple of Kukulkan, Yukata, Mexico | Salle des Cariatides, Louvre, Paris, France

Kleiner Tornowsee (Little Tornow pond) leans towards the shadows. 'It is as if these dark waters have a special pull into the depths and as if they stand more rigid and immobile than others,' wrote the German poet Theodor Fontane in the nineteenth century. Midway through the Industrial Age Fontane walked the pastoral landscapes around Berlin. His aim: to collect the little marvels and sentiments that would arise here and there within and without. Kleiner Tornowsee was one of them.

It is quiet here. No sound of birds, no chirping insects. The silence of the glaciers that once carved the landscape seems now embedded in the pond. They still seem to reverberate into this hollow trough, a shudder and tremble among the oaks, alders, ashes. Even today, the wanderer has to divert off the well-trodden path, find their way past the riparian zone, then slip under the veil of a presence that hovers over the surface of the pond.

It is easy here to imagine a world long gone. Fontane writes of nymphs surfacing on midsummer day, a saddened twinkle in their eye. If ever they existed, they have left this world, reached another shore, moved behind another veil.

On the southern bank a weathered sign reads *Echostein*, echo stone. Underneath, hardly noticeable, a grey rock to stand on. No further instructions, just this setting: a sign, a stone, the allure of an echo. 'It really works!' The comments left on outdoor tracking sites. Indeed, every word spoken out loud, with feet resting heavy on the stone and directed across the waters, will come back, syllable by syllable, a hollow voice birthed from one's own.

There are lists for everything on the Internet. Best locations with echoes enumerate mausoleums, museums, temples. Gol Gumbaz mausoleum of Bijapur, India. El Castillo Temple of Kukulkan in Yukata, Mexico. The Salle des Cariatides in the Louvre, Paris. Buildings to enshrine past collective memories. Buildings to preserve what has long gone. Is the echo not a ghost, zigzagging its way through tunnels and empty chambers, filling the voids of architecture? A flux of movement, a sudden tremor in the make-up of this world, linking experiences and narratives.

The science is more precise. A cast sound bounces off a surface, not unlike a reflection on still water. However, sound waves travel slower than light waves and a delay occurs. It is the distance between the sound's source, the listener and the reflecting surface that will generate the echo.

'The ancients built Valdrada on the shores of a lake, with houses all verandas one above the other, and high streets whose railed parapets look out over the water. Thus the traveller, arriving, sees two cities: one erect above the lake, and the other reflected, upside down.' After evoking this place to a fictional Kublai Khan in his *Invisible Cities*, Italo Calvino sets a pause. The Khan contemplates. In turn he describes a town to Marco Polo, a place he saw in dreams, a luring from a different shore. He orders Marco Polo to set out and seek this city: 'Then come back and tell me if my dream corresponds to reality.' 'Forgive me, my lord, there is no doubt that sooner or later I shall set sail from that dock,' says Polo, 'but I shall not come back to tell you about it. The city exists and it has a simple secret: it knows only departures, not returns.'

The *kodama* are tree spirits in Japanese lore. One hears them at the dead of night, the sound of a falling tree when no tree falls, a breaking twig when no one is around to break the twig. Sometimes the echo sounds in a valley, a gorge, a cliff. They are as ancient as the islands. When *kodama* take the shape of a *yamabiko*, an echo child, they can fall in love with humans.

'It is sound that lives in her,' wrote Roman poet Ovid in the *Metamorphoses* about Echo, the nymph who longed to unite with self-absorbed Narcissus. Her life was not the same after Hera, wife of Zeus, cursed the girl to only repeat the last word of anything spoken to her. Hence, she could not reach out to her beloved: 'and when he said "Goodbye!" Echo also said "Goodbye!"' After Narcissus' death, Echo withered away, became a stone, until only her voice was left, haunting the silence.

'Ah, what an age it is/ When to speak of trees is almost a crime/ For it is a kind of silence about injustice!' So wrote German playwright Bertolt Brecht at the end of the Second World War in his poem 'To Posterity'. Some time after the war, Brecht moved to the little town of Buckow, seeking refuge from the traumas he had witnessed. It is here where the hiking path to Kleiner Tornowsee begins. How often must Brecht have sought out its shadows? What words did he choose to whisper across the languid surface of the water and which answers did he get, standing steadfast on the stone? What silences surround us now?

17

Sounds of the Atomic Age

White Sands Missile Range, New Mexico, USA | Yucca Flat, Nevada Proving Grounds, USA | Rongelap Atoll, Marshall Islands | Moruroa and Fangataufa Atolls, French Polynesia

16 July 1945. 5:29 a.m. The weapon codenamed 'Trinity' is successfully detonated at the White Sands Missile Range in New Mexico. The heat from the ensuing blast, from 16 kilometres (10 mi.) away, feels like opening up an oven door. A troop transport flying west describes the sight as a second sun rising from the south. Beneath, the desert sand fuses into a never before seen green-glass rock later named 'tritinite'.

In his official report, Brigadier General Thomas F. Farrell focuses on the quality of light: 'It was golden, purple, violet, gray and blue. It lighted every peak, crevasse and ridge of the nearby mountain range with a clarity and beauty that cannot be described but must be seen to be imagined.' The mushroom cloud would mark humanity's entry into the Atomic Age.

What is less iconic than the sight of a nuclear explosion is the sound it makes. 'A Noiseless Flash' – the title of the opening chapter of reporter John Hersey's collected accounts of Hiroshima survivors, published in the *New Yorker* in 1946. 'Almost no one recalls hearing the noise of the bomb,' Hersey notes. Most likely the sheer force, heat and amount of debris was enough to overwhelm anyone lucky (or unlucky) enough to survive.

For a sufficiently large explosion viewed from a sufficiently safe distance, there will always be a lag between image and sound. Popular media depictions and reconstructions tend to compress both into a single monstrous phenomenon. Notes by Robert L. Mott on crafting the sounds of a nuclear blast in his seminal book, *Sound Effects: Radio, TV, and Film*: a building being dynamited to mirror the initial blast, followed by the slowed-down roar of a waterfall for the sustain and decay phases.

When mixed correctly, this lie can become authentic and believable. 'Not one viewer, not one critic, not even the scientists who actually developed the bomb complained about this odd mixture of sounds when they appeared on the TV news in the early 1950s,' Mott boasts. 'What everyone heard matched convincingly the picture.'

What, then, is the truth? In an effort to allay public fears on nuclear weapons testing, civilian reporters were permitted to observe 'Annie', one of eleven weapons tests conducted for Project Upshot–Knothole at Yucca Flat at the Nevada Proving Grounds. The unedited audio recorded from the test, coming more than 30 seconds after the first flash, is not the dull and rising roar one might expect, but more of a sharp rifle crack, shortly followed by an expletive from an unknown soldier.

Excerpts from the media interview afterwards with a pleased General John R. Hodge: 'The test, I think, went very well.' On the reaction of his troops witnessing the blast equating to 16,000 tons of TNT. 'They took it in stride, as American soldiers take all things.' The American public would take the full brunt in the end in the form of radioiodine exposure; the National Cancer Institute estimated that Project Upshot–Knothole was responsible for 28,000 additional cases of thyroid cancer and 1,400 deaths.

Nuclear fallout makes no sound, can fall as light as pure snow. Before their skins peeled and their bodies grew tumours, the children on Rongelap Atoll in the Marshall Islands played in a never before seen coral ash that drifted down on their shores, the result of Operation Castle Bravo, the 1954 nuclear test at Bikini Atoll. The hydrogen bomb was three times stronger than calculated and a thousand times stronger than the one dropped on Hiroshima. More weapons tests by more nuclear powers on isles across the South Pacific over the next five decades: the United Kingdom with Operation Grapple on Kiritimati and Malden Island; the French later on Moruroa and Fangataufa atolls. The latter's 1974 test, code-named 'Centaur', would expose the entire population of Tahiti to dangerous levels of radiation when the resulting cloud took an unexpected trajectory.

There is a lag between act and admission. There is an even longer lag between culpability and compensation. Sixty years on, the former children of Bikini Atoll are now exiled elders, still awaiting restitution that may never come, unable to return to their now poisoned paradise, fearful always of sounds that do not ever seem to fade: the crackle-click warnings of a Geiger counter – perhaps the true enduring signal of the Atomic Age.

18

Sounds from the Deepest Artificial Point on Earth

Kola Borehole, Kola Peninsula, Russia | Stratavator, Pittsburgh Museum of Natural History, USA

Even in a horizontal, flat landscape, cosmology runs vertical. From up – heaven, to down – hell. In between – that thin crust upon which life is rooted. A tender, fragile layer that humans turn into infernal landscapes.

Vilgiskoddeoayvinyarvi, the Wolf Lake on the Mountains, was once home to the indigenous Sami people and their livestock. Now its lakes, rivers and swamps leak poison. Copper and nickel mines and smelters have turned the sparse soils upside down. Only the wind and some birds sweep over the hills; most other life has faded away on the subarctic Kola Peninsula, where a border separates Russia from Norway.

Somewhere in these barren lands, a ruined industrial tower rises, its debris strewn across the ground. At the foot of the tower, a metal lid of sorts, an arm's length in diameter, welded and sealed with metal screws, thick, heavy, rusted. Underneath: the entrance to the deepest artificial hole on Earth.

'The world is deep: and deeper than day has ever comprehended,' writes Nietzsche in *Thus Spoke Zarathustra*. Drillings on the Kola Borehole started on 24 May 1970 as part of the former Soviet Union's programme 'Investigation of the Continental Crust by Means of Deep Drilling'. That very year Lenin would have celebrated his one-hundredth birthday and the committee was eager to support projects that boosted Russia in the Cold War: this time, by exploring the solid darkness below rather than the infinite night above. Promises of knowledge about the make-up of Earth. Understandings of plate tectonics. More precise locating for metals, minerals. Maybe a clue about Hollow Earth, even? The idea was to get one-third of the way through the Baltic Shield continental crust on Kola, roughly 35 kilometres (11 mi.) thick. Another 6,000 kilometres (1,800 mi.) before the core of the planet. Still, it would be deeper than anyone had ever dared, even the Americans.

In 1984 Soviet scientists had reached 12.26 kilometres (7½ mi.) and celebrated by inviting experts from around the world. The three-day excursion to the Kola Borehole included a coach tour of Murmansk, a concert at Oktyabr Community Center and a farewell party. As a present to the guests: coffee cups with the image of the iconic drilling tower.

The act of drilling is an exploration of time as told in layers. At 1 kilometre (½ mi.) deep the scientists found magnetite, copper and nickel, water. At about 3 kilometres (2 mi.) deep they discovered rock that was similar to the samples from rocks carried back from the moon (alas, by the Americans). At 10 kilometres (6 mi.) they hit rocks 2.5 billion years old, saturated with microscopic plankton fossils. Down there temperatures rose to 180°C (356°F) and quickly climbed the deeper they went. Stones turned ductile and viscous. Drills lost footing in a molasses-like substance. When no technology could allow them to go further, the scientists started to listen.

In the 1990s a recording from the borehole made its rounds: first, screams and yells. Christian papers in the United States readily accepted them to be the voices of tormented humans. What did the Russians really discover in that unfathomable deep hole? Had they opened up hell?

It turned out to be a hoax. The infernal soundtrack was nothing but a soundscape sampled from *Baron Blood*, a 1970s Italian horror movie, mixed with the rumblings of the New York subway. But it fired the imaginations of many believers long before fake news became commonplace.

The real recordings from the hole were made with sound sensors and geophones that tracked seismic vibrations. At first there was a roaring. Then it stopped. It started again the next day. And stopped. It took some deep listening before the scientists could link the vibrations with the working hours of a nearby copper mine. The earth was feeding them man-made sounds. They kept their sensors steady, tuned. Other signals became audible. The longer they listened, the more they discovered a soundscape of echoes and crunches, of vibrations and ultrasonic sounds. What were the rhythms, and what did they speak of?

Many years later sound artist Justin Bennett visited the Kola Borehole to record the mood of the site. He interviewed Yuri Smirnov, once chief geologist of the project, who still lived nearby in a room stuffed with probes from deep within the earth, probes that felt cold when held against the skin. Smirnov wrote poetry, was familiar with Dante's *Inferno*, which he found fascinating but became bored of its morals and politics. He loved to display the medals he was awarded during his almost forty working years at the site, of which the last ten were spent doing nothing much as funding had been cut. The project finally closed in 2007.

For his installation 'Wolf Lake on the Mountain', Bennett created a fictional character called Victor Koslovsky, modelled after Smirnov. He imagines the man revisiting the site regularly, looking out at the landscape, watching the birds circling around the borehole, musing about this acupuncture point of the earth. Koslovsky waxes lyrical about how people have forgotten to listen. How he is slowly learning to predict the future by focusing on the shifting vibrations of the earth. And how the open borehole sounds like a trumpet, with the wind blowing in.

The Pittsburgh Natural History Museum has a stratavator, an elevator that simulates a ride down to Earth's core, where visitors hear the sounds of wind trapped in the shaft and the cranking of the industrial elevator. 'This is as far as we go,' the excited miner on a video screen explains when the ride stops at 5,000 metres (16,400 ft). Since the closing of the Kola Borehole other places have started to dig deep. In Qatar the record now

Sounds from the Deepest Artificial Point on Earth

has been broken for the deepest hole on Earth. No one listens into the Al Shaheen site. Here they drill for oil, bringing up more fossil fuels that will be burnt. This is as far as we go.

19

The Humming Fields and Meadows of the Altai

Altai Republic, Russia

Imagine a country where everyone lives in peace and harmony. A land not marked on any map, not even outlined as a white spot. Hidden, somewhere in the mountains, far, far away. Only to be found by those who actively seek.

One day at the turn of the tenth century, Vladimir the Great, Grand Prince of Kiev and ruler of Kievan Rus, consulted with a travelling monk about where such a land could be found. The monk entered a state of trance: 'Search far far away, in the east. But be aware. Only those who truly seek will find it.'

To reach the mountains of the Altai today, one needs to cross the Siberian steppe for six days and nights on Europop-blasting trains and bumpy buses. Located almost at the centre of the Eurasian continent, an ocean of grass, the dark taiga forests or the arid lands of Kazakhstan and Mongolia lie between here and the rest of the world.

The Humming Fields and Meadows of the Altai

In the Altai, industrialization is but a bad rumour from where people keep busy with too much stuff. Here, habitats have been made of wood or stone, clay or leather, and decompose faster than the span of a human life. No castles, fortresses, temples. No cities, airports, train tracks. No factories. No nation-building before annexation by the Russian state. History is but a vague concept in a landscape where nomads drift in and out like clouds, winds and swallows, the last of which they sing songs for at the end of each summer. Come back again, they yearn, and celebrate their kinship, the sharing of air, water, soils, with immaterial praise. A song. No more.

The world is always in the making. Was it a hundred, a thousand, or ten thousand years ago that a wanderer carved 'A lost one, one who seeks the true path, has written this' into the stones near the Katun river, a silver glistening band? Ten thousand years ago, a sign confirms at the petroglyphs along the Tschuiski trakt, the only highway in a land of unpaved roads. Not much, it seems, has changed in what civilization would call a myria-annum. The bellows of the river down below, the humming of the insects in the fields above, maybe the cry of a bird.

In summer, the Siberian sun turns every rock or blade of grass the lustre of Russian icons. Sometimes a dirt road branches away from the Tschuiski trakt. If one follows it for a while, the soundscape distances itself from the GAZ motors and falls back into the seasons. Time does not stand still in this pre-industrial landscape. But its rhythm is ancient and the voices heard tell of this.

Those who have sought to come, stop and close their eyes. They stretch out their hands, combing through the chest-high grass and flowers, breathing in the tangy notes of orchids and irises, spruce and edelweiss. Six hundred species of mosses, 1,200 lichens, over 3,500 species of vascular plants. And hovering above and crawling beneath and spinning webs in between, insects, birds, bees. The hum, the buzz, the audible interplay of billions of creatures.

Many describe this as the true essence of the place, the humming meadows of the Altai. It is a sound so calming and soothing that some say it transports the mind and spirit to a portal, possibly the entrance to that forgotten place, where everyone – humans, non-humans, animated, non-animated – can live in peace and harmony. Did every place along this latitude once sound like this?

> And so we're hearing the tangle of all these narratives. I think partly it's a delightful break from the particular human way of thinking and of listening, but more broadly I think it teaches us that this is the challenge of our time: how do competing narratives, thousands of them, find their way to one another so that they intersect without bringing the whole edifice down?

So asks David Haskell, author of *Sounds Wild and Broken*, referring to the thrums of the natural world, of thousands of species cohabiting in one space in unison. 'I think it gives us an appreciation for the benefits of anarchy; not anarchy as a destructive force, but anarchy in the sense that there is no governing central authority, as there is, and as there should be, in human music.' It does something to the psyche, listening to this.

When Russians travel here to one of the newly established eco-lodges, their friends back home will ask them to bring back honey. Just a jar of honey. And in it the distilled spirit of summer, the interplay of insects, flowers, the soil beneath, the weather, the sun.

Only now sometimes snow falls in summer. Light white flakes, sometimes mixed with chunks of metal debris. Baikonur, the world's biggest space port, lies beyond the mountains in Kazakhstan. Along the rockets' trajectory is a path of death due to UDMH, a toxic rocket propellant. Cancer and strokes befall humans and animals alike. While some still search for the land of peace and harmony in the worlds and stars above, one of the last paradises on Earth lies vulnerable, at our mercy.

20

Otherworldly Ordinary: The Found Sounds of the Fantastic

Edmonds Lake, Anchorage, Alaska, USA | Lake Baikal, South Siberia, Russia | Steamboat Lake, Colorado, USA | Monument Valley, Utah, USA | Pocono Mountains, Pennsylvania, USA

When incongruities arise between the visual and sonic worlds, we may not be able to square what we hear with what we see. When Cory Williams skipped a rock across the frozen surface of Edmonds Lake in Anchorage, Alaska, he did not expect to hear the scatter of laser gunfire, or that his YouTube video which captured his astonishment would go viral. Subsequent instances of otherworldly emanations from fracturing lakes: photographer Alexey Kolganov while skating over Lake Baikal, where bursts more resembled cannons than pistols, booming and resonant, possibly amplified by the world's deepest lake; the Colorado Parks and Wildlife department's social media account, with samples taken from Steamboat Lake that eerily mirror the iconic blaster sounds found in the original *Star Wars*, known so well by generations of movie-goers.

Such otherworldly sounds from earthbound sources naturally foster intrigue, demand explanation. Conspiracy theorists are first to seize upon the immediate and most far-fetched of conclusions. 'This is 100 per cent proof of an alien base below this lake,' one asserts, convinced at the presence of extraterrestrial labs testing clandestine weapons. Here incongruities arise once again, this time between reality and fantasy. Real-world lasers may clack and whirl from their power supplies and cooling systems, but the beam of light itself is silent. Physicists, while slower than conspiracy theorists at drawing their explanations, arrive at sturdier, more provable ones. They attribute the cause of these lake sounds to the phenomenon of acoustic dispersion, where frequencies travelling at different speeds through the ice separate out. Thus the high trills reach our ears first, followed by the lower spectral booms. A strange but simple phenomenon, easily unravelled.

There are those who would travel far for strange sounds to tell strange stories. Composer Hans Zimmer journeyed into the deserts of Monument Valley to seek sonic inspiration for the 2021 film adaptation of *Dune*, directed by Denis Villeneuve. 'How does the wind howl through the rocks?' Zimmer muses as he struggles to convey the ambience of the film's arid world. 'How does the sand grit in your teeth?' His crew buried hydrophones in the ground in an attempt to understand how the movie's giant sandworms would sound travelling beneath the dunes, reimagined as real whales swimming beneath the seas.

Sand sounds as surf sounds. Desert depths as ocean abyss. A transformative auditory exercise, a means to breathe life into fantastical beasts.

Yet most of the time, the most extraordinary sounds come from the most mundane activities. No far-off lake ice played a role in rendering the genuine *Star Wars* blaster sound. Instead it came to renowned sound designer Ben Burtt while backpacking with his family in the Pocono Mountains of northeastern Pennsylvania: the guy wire of an AM radio tower, struck with a hammer and recorded with a microphone. The rest is history.

This is the art and skill of the Foley artist, aligning what audiences hear with what they see on screen, matching their expectations to their imaginations. So the clacking of two coconut halves together becomes the trotting of a horse; the snapping of celery sticks becomes the breaking of bone.

Close your eyes. There is magic to be heard everywhere. The bacon frying in the pan is a passport to a landscape of falling rain. Crinkle some cellophane wrap to return home to a cozy fire. Even distant travel is possible at a whim through found sounds. Stretch out a metal slinky toy between two points and strike one part. A home-made version of striking a hammer against a guy wire. Maybe after reading this you will be transported to the shores of frozen lakes. But more likely you will find yourself yearning for adventure in a galaxy far, far away.

21

Pay Attention: On the Sounds of the In-Between

'The Microsoft Sound', Windows 95 | 'Opus No. 1', Cisco Call Manager Default Hold Music | *Hassha merodii*, Japanese train departure melodies | *Ambient 1: Music for Airports,* by Brian Eno

More than a decade before the launch of the first iPhone, tens of thousands would line up at midnight in anticipation of a new future delivered through consumer technology. Windows 95: the operating system that promised to usher in the burgeoning Internet Age. Before the dial tone and hiss of the modem handshake took users into the World Wide Web, they would first be greeted by Window 95's start-up chime, later simply known as 'The Microsoft Sound'.

Famed British sound artist Brian Eno was approached by the software giant to produce the iconic jingle. It should be inspiring and universal, Eno recalled the instructions he was given, optimistic and futuristic. Also emotional yet sentimental, yes.

Oh, and it has to be three and a quarter seconds long. Eno threw himself into the task, eventually coming up with 84 versions – 83 of which have yet to see the light of day. 'It's like making a tiny little jewel.' The musician reflected on that final arpeggio, dreamlike and ethereal, forever cemented in sonic history.

There would be other tunes for other operating systems. The surround sound showcase of Windows 98. The fresh breezy stylings of Windows XP. In recent years startup sounds have grown shorter, less evocative and more minimal. Generic. No place for cheery greetings in an era of always-on tablets and phones, used everywhere, anytime. Still, Jensen Harris, former director of the Windows User Experience team, expressed regret at being the one to axe the theme.

For an age with seemingly no time for small welcomes, a surprisingly large amount of time is wasted in limbo. Tech giant Cisco, having sold over 100 million corporate phones, has played no small role in putting the world on hold. This all too familiar experience might be even more intolerable if not for the Call Manager Default Hold music. Originally titled 'Opus No. 1', the piece was composed by a sixteen-year-old Tim Carleton in 1989 and recorded on four track by his high-school friend Darrick Deel, who went on to work at Cisco designing phone systems. With Carleton's permission, Deel installed the five-minute piece of tape on to the default setting.

No radio feature or top ten list was needed for the stirring synths of Opus No. 1 to reach millions and attain cult earworm status. 'Perfect music for a night drive.' 'Play this at my wedding one day.' 'The only on-hold music that made me want to stay on hold.' For many, Carleton's teenage musical experiment became an antidote to the barrage of 'we are currently experiencing a high call volume', another gem that preserves one's sanity in the struggle against Kafkaesque bureaucracy.

Small sounds to smooth the edges of modern life still require a human touch, it seems. Perhaps there is no one who believes this more than Minoru Mukaiya, former keyboardist for the legendary 1980s jazz fusion band Casiopea, and current composer of *hassha merodii* – departure train melodies – for stations across Japan. Each of Mukaiya's seven-second compositions takes into account how the train enters the station, the way the tracks curve, where it will be going. The rising melody at Shibuya Station is uplifting to reflect the climb towards the next platform. The tune at Monzen-Nakacho uses traditional instruments to honour the neighbourhood's older demographic. The theme of *Astro Boy* plays at Takadanobaba, a nod to it being the setting for the iconic manga series. Every one of the 170 jingles allows passengers 'dwell time' for boarding; as riders grow more familiar with the melodies, they learn to gauge whether they have enough time to make the train, becoming less inclined to rush and stress. Mukaiya insists on playing each piece himself by hand, believing that the imperfections of 'the human groove' render them more intimate and joyful. Fans have told the jingle maker that his ditties help to warm them up after a hard day's commute. Japan has 44 of the top fifty busiest train stations in the world. Day in and day out, an ethos of care in so small a thing seems to matter a great deal to many.

What is the soundtrack of the interstitial? What is the music of the interlude, between act and act? Inspiration found the aforementioned Eno again, this time on a Sunday morning while waiting at Cologne Bonn Airport. Entranced by the beauty of the architecture but offended by the selection of music, Eno went on to create what is now known as ambient music as a means to explore the acoustic and atmospheric quirks of public spaces. Released in 1978, *Ambient 1: Music for Airports* became a powerful counter to the canned elevator muzak of the day, and was intended to 'induce calm and create a space to think'. From the album liner note: 'Ambient Music must be able to accommodate many levels of listening attention without enforcing one in particular; it must be as ignorable as it is interesting.' On the latter objective, Eno has not been entirely successful. More than four decades after its release, *Music for Airports* has drawn attention and accolades as a pioneering masterpiece, growing too compelling to become background. 'The music is coy,' writes John T. Lysaker, professor of philosophy at Emory University:

> It catches our attention and then leaves it, suspended. Sometimes, we're just caught between hearing and listening. Other times reverie results. Whatever we were doing becomes slightly strange, something to think more about, perhaps pursue differently. The album thus accomplishes

something that a lot of powerful art accomplishes. It dislodges the familiar. But rather than replace it with some new vision, it leaves us to ponder what might yet be.

Perhaps this is what sounds of the in-between should strive to be, not to command our attention directly, but rather allow us space to unfold into the world, joining together what we have done and what we will do with the present moment we are occupying, here in the now.

22

The Not-So-Secret Love Lives of Fish

Sausalito, California, USA | Cape Coral, Florida, USA | The Colorado River Delta, Gulf of California, Mexico

One expects creature cacophonies from frogs and birds and cicadas, but fish?

Houseboaters living in Sausalito, California, during the early 1980s were treated to the love songs of the humming toadfish (*Porichthys notatus*). Also known as the plainfin midshipman, their mating thrums were described as resembling a room full of oboes, the drones of B-52 bombers, or live electric razors being dropped into water. Initial theories around the origins of this mystery sound before it was pinned down to the 20-centimetre (8 in.) fish included: secret submarine projects; sound-leaking electric cables; discharges from the nearby sewage plant; nerve gas being pumped into the bay by the CIA. Houseboaters with bedrooms at or below the waterline suffered the most, as the calls amplified across reinforced concrete hulls. Earplugs proved useless. 'As the night gets quieter, the humming gets louder,' a Ms Suzanne Dunwell remarked. 'It may be the mating call of the toadfish, but it plays havoc with the sex lives of people living here!'

Embracing the hand they were dealt, residents of the bay town began hosting an annual Humming Toadfish Festival. Revellers dressed up as sea monsters and dancing fish to parade around the dock, listening to kazoo versions of 'Yes, We Have No Bananas' and 'The Battle Hymn of the Republic'. There was even the crowning of a Toadfish King. 'The toadfish is the perfect thing [for this town],' one resident and festival attendee remarked. 'It's eccentric, it's peculiar, it's off the wall.' Only in Sausalito, it seems – at least until the toadfish found new breeding grounds a few years later.

The year is 2005. Growing tired of having their peaceable evenings ruined, retirees wintering in Cape Coral, Florida, blamed the local utility company for nightly beats that began each evening near dusk and continued several hours past nightfall. When the cause was finally discovered, some complainants were in disbelief: 'There was no way a fish could produce a sound that could be heard inside a house.' The man-made channels of the housing development proved the perfect habitat for the black drum (*Pogonias cromis*) to hide from sharks while they serenade away. Capable of living sixty years and growing to 45 kilograms (100 lb), males flex abdominal muscles against swim bladders to produce courtship overtures in the 100 to 500 Hz range – low and deep enough to carry across canal walls and into the living rooms of prime waterfront real estate.

THE SOUND ATLAS

The world record for loudest fish belongs to the Gulf corvina (*Cynoscion othonopterus*), a black drum relative endemic to the upper Gulf of California, Mexico. Each spring, the entire adult population congregates in the mouth of the Colorado River Delta to mate; estimates of up to 1.5 million fish along a 27-kilometre (17 mi.) stretch were recorded one peak spawning day. At 177 decibels, the machine-gun fire romancing call of a male corvina underwater is as loud as a rock concert on land, louder than standing next to a chainsaw – just below the human threshold for pain. Dolphins and California sea lions feeding around the area risk temporary or even permanent hearing loss under such sound exposure levels. Researchers passing over a spawning cluster reported a low hum vibrating through the hull of their boat, which proceeded to grow in volume to resemble a gigantic swarm of bees. Despite being out of the water, they had to shout at each other to be heard.

Gulf corvinas may have adapted to the Lombard reflex, escalating their call volume to better communicate in an inherently noisy and turbid environment. They may also have resolved the cocktail party problem, seemingly able to discern individual calls out of the collective chorus. This vocal prowess combined with their gregarious nature proves both a blessing and a curse. Homing in on their raucous calls, a single panga boat can net up to 2 tons of fish within minutes; the local fleet of five hundred can take in 2 million fish in less than three weeks – the species is currently listed as 'vulnerable'.

23

Voices Within and Without

Pyramid of Merkine, Lithuania

When seven-year-old Povilas Zekas went to church together with his grandmother on 19 August 1990, a voice spoke to him. It was bodiless, sweet in tone, and introduced itself as his guardian angel. From then on, Povilas would hear it often.

Ten per cent of the world population claim to have been contacted by unknown entities. One-third of those report it to be a terrifying experience: strange utterances and screeches, unresolved trauma welling up from within. 'Cries of tigers and hyenas' – the composer Robert Schumann ended his life in an asylum. The other two-thirds are comforted by *kol demamah dakah* – 'the sound of a slender silence', as it translates from old Hebrew in the Bible. After enduring wind, earthquake and fire, the prophet Elijah finally hears a voice after which no doubt remains in his heart. He wraps his face in a mantle, leaves his cave behind and stands under the open sky to listen. Elijah was neither the first nor the last. Martin Luther King fell asleep on his kitchen table and woke up to words that took his fears away forever. Philip K. Dick found the messages he received from his 'no-body' economical in their cadence – clear instructions phrased in the least amount of words. For Gandhi and Joan of Arc, the voice of God helped them shape a new world. Socrates was simply told to stay out of politics.

After their first acoustic encounter, Povilas's angel revealed itself in the evening sky. A ray of light, golden and warm, touched the meadows below. A cool breeze was sweeping in from the forest nearby; the marsh frogs were croaking. Povilas stood on the porch of the house with his grandmother, and together they watched in awe. Povilas would build a pyramid on this exact spot when he turned nineteen. Made from aluminium, not steel – the voice was clear about this. Later, a geodesic dome was needed to shelter the pyramid. Povilas, then 26, received the message and again abided. He never doubted the voice. Its source came from a place that was neither inside nor outside his body.

The controversial theory of the bicameral mind goes like this: one side of the brain was used to speak while the other listened and obeyed. The *Iliad* is a story reputedly driven by this power play between the lobes. Ancient heroes heard messages and acted upon them. Talking back and forth between the self and the gods is now a thing of the past, the theory states. We have evolved.

Povilas's dome stands in the middle of a national park. The closest hamlet has five registered residents. In summer, the sounds of nature are a celebration of life: chirping, trilling, lisping and buzzing; the susurrus among the trees. Inside the dome is silence. Any sound struck – a raising of one's voice, a gentle touching of a harp's string, a gale of breath into a flute – will reverberate against the triangular-shaped glass ceiling patterns and transform into melodic echoes, as celestial harmonies to be experienced on Earth. People travel from far and wide to listen. Povilas must have gotten something right in the construction plans that he downloaded from his guardian angel.

When Povilas was a little boy, the Singing Revolution triggered the events that led to independence from the Soviet Union. The winds of change came through music. Holding hands and joining together in folk songs from the past, the Baltic people stood up for freedom. In the aftermath, New Age religions sprouted up in Lithuania like mushrooms in the autumn forests. They filled the vacuum the Socialist regime had left.

Today Povilas spends his days guiding visitors into the dome. More than 50,000 tourists come each year to these southern parts of the old new country. Povilas will retell his story with a kind face and then start singing with an even kinder face. He will chant away, sending sound waves manifold into the space within the dome's structure and provide a summer-rain of music. The sound is said to heal diseases, enabling people in wheelchairs to get up and walk or making unbelievers realize the presence beyond. Visitors may experience their first hearing of their inner voice, right here. If not, there are other ways to learn.

Everyone has the ability 'to treat what the mind imagines as more real than the world one knows', writes Stanford anthropological psychologist Tanya M. Luhrmann in her book *When God Talks Back*. Hearing God's voice is a matter of training perception, she muses, not unlike gaining the body awareness of a tennis player, honing the delicate palate of a sommelier, or fine-tuning the eye of a sonogram technician to distinguish shades of grey in a single organic, trembling snapshot.

24

Listening to Traces

Embankment station, London, UK | Wind phone, Otsuchi, Japan

Embankment station in London is made up of many layers. The rumble one might experience above on the street is likely to stem from the original subsurface lines and the deep-level tube lines underneath. Every so often Dr Margaret McCollum will go down deep into the tunnels, find a bench on the platform of the northbound Northern Line, and sit down until, accompanied by a gust of wind, the next train approaches. When she hears the 'Mind the Gap' announcement, a smile lights up her face.

Tourists in London have always enjoyed the very British rendering of the phrase, but the one in Embankment sounds especially like the Queen's English, as if spoken by a character in the television series *Downton Abbey* (2010–15): 'Mind' – [short pause] – 'the Gap' – leaving another audible pause at the end. The timbre of the voice gives a gentle warning. The slight tremble reminds listeners of an era when sending passengers with tight schedules on invisible underground networks was a technological achievement.

The voice belonged to Oswald Laurence, trained at the Royal Academy of Dramatic Arts, who half a century ago recorded the announcement and was once heard on the entire Northern Line. Dr Margaret McCollum is his widow. When the old announcement was exchanged for a more modern rendering, she politely asked the authorities to put her late husband back on again. The authorities agreed.

When not opting for the ever-more-common computer-generated voice announcements, some cities seek a certain nuance in their choice of announcers. For a long time on the subway in Berlin, it was voice actress Ingrid Metz, semi-famous as the dubbed German voice of Marilyn Monroe and luscious 1980s chocolate praline adverts, who proclaimed the halts. A gasped 'Kottbusser Tor' or 'next stop Grunewald' would send shivers down the spine of many a commuter, giving the passengers a thrill of excitement before they re-emerged into the world outside. Since December 2020, Metz's voice has been replaced by Philippa Jarke, who offers a transgender nuance to the city's diverse places.

One could easily take a detour here to the fandom of train announcements, referring for example to Arriva trains in Wales, and their announcement listing 31 stations in both English and Welsh that takes 2 minutes 45 seconds to say:

Listening to Traces

The train now arriving at platform one is the 15.57 to Crewe stopping at Swansea, Gowerton, Llanelli, Bynea, Llangadog, Llanwrada, Llandovery, Cynghordy, Sugar Loaf, Llanwrtyd, Llangammarch, Garth, Calmer, Builth Road, Llandrindod, Pen-y-bont, Dolau, Llanbister Road, Llangunllo, Knucklas, Knighton, Bucknell, Hopton Heath, Broome, Craven Arms, Church Stretton, Shrewsbury, Wern, Prees, Wrenbury and Nantwich before arriving in Crewe.

It is reported that many a commuter misses their next train in their attempt to listen to this tongue-twisting experience. But this is not the scope of this particular sound excursion, and so we stop here and ask anyone interested in learning more about this topic to leave this train of thought and find their path in the wild terrains of the Internet.

~~~

What do we hear in a voice? What does a certain tremble in the speaker tell us about the layers of experience they have had in their lives? We expect a singer to guide us through the lyrics of a song while at the same time to fill the words with lived emotion. A quiver in a voice can open portholes to deeper layers of a soul. Sometimes all that is left of a voice are its traces kept within a recording or nature itself. Starlings mimic their surroundings, including everything from the farm machines, whistles and human voices to the background ambience of their habitat. For the Irish writer Doireann Ní Ghríofa, these environmental sounds are absorbed over generations of starlings and woven into the fabric of their songs:

> Oftentimes, what I find very moving is in Ireland, you'll often happen upon a little cluster of ruined houses that may have been abandoned during the famine in the nineteenth century and there will always be starlings around. I find it intensely moving to listen to the starlings there because over generations, they learn and internalize the sounds that are there. You know when they will be mimicking sounds that were made by people who were long gone from those dwellings.

It is as if the ghosts and spirits of those who once dwelt there are still alive in the starlings' trills.

Sometimes it is the echo of nothing that makes us listen for the voice we long for, and by doing so, we become aware of something else. Writer Maggie Jones recounts in the *New York Times* how she kept calling her mother for months after she had passed away. At first she thought it was to listen to the recording of her voice message, which no one had bothered to disconnect. But then she began to tell stories to the voice; things, happenings after a long day, a fight with someone, just as she had done when her mother had still been around. She did not mind that no one spoke back; simply hearing the greeting triggered a deeper layer of emotional response within. 'Eventually I realized there was a pattern to my messages: They often reflected how I thought my mom would reply to me or the advice she would give.' Jones had learned to internalize the guidance of her mother, just as she had received it all her life.

~~~

No traces are left of the 20,000 people who were wiped off the Earth's surface during the Japanese tsunami of 2011; their homes and daily routines disappeared with the waters. For their lovers, friends and relatives, there was no way to say goodbye. Where there are no visible traces left, it can help to create a placeholder. The *Kaze no Denwa*, the wind phone, was constructed one year before the disaster by landscape designer Itaru Sasaki: a white phone booth in a quaint, windswept garden for the memory of his deceased cousin. 'Because my thoughts couldn't be relayed over a regular phone line, I wanted them to be carried on the wind.' The *Kaze no Denwa* has since been visited by 30,000 survivors longing to connect. While the phone stays silent, they sob their messages into the receiver. The wind carries them to where they need to go.

25

Explaining the Inexplicable: The Taos Hum

Taos, New Mexico, USA

Scents of juniper and mugwort bushes. Scattered splinters and shards of rock storing daylight heat, radiating it long into the night. At dusk a cold breeze mixes with the fragrant air, causing a drift that tickles the fine needles of nearby spruces. The stars come out, stretching across the sky to the mysterious beyond. A last shudder over the landscape. A blanket of silence. Then, the Hum sets in.

'It sounds like when there's a DVD in the player and it's spinning around, but the TV isn't on.' 'It sounds like the deep hum of a distant irregular diesel generator with intensity fluctuations.' 'Low frequency sound not able to locate. When it starts only moving my body or turning my head stops it. It mostly is recognized when awake in bed and comes for some hours. Sometimes several nights, sometimes not recognized for months.' A sampling of comments from the World Hum

Explaining the Inexplicable

Map describing the sound of the Hum, an acoustic phenomenon that started in Taos, New Mexico, in 1991. It is louder at night and usually sets in between 10 and 10.30 p.m. Roughly 2 per cent of the local population experience it. The Hum stops when one moves and is usually heard inside one's body.

Born in 1801 in Lombardia, Italy, Juan Maria de Agostini left his home as a young man to become a hermit in the American west. On his way south he joined a caravan and despite being offered a seat on one of the wagons, chose to walk all the way to Taos, New Mexico. At the Sangre de Cristo (Blood of Christ) Mountains, named after the light of the evening sun reflecting on its slopes, he stopped and stayed. One of its many caves became Agostini's new home. He listened to the waterfalls cascading down the mountain, a musical score of seven different tones, enhanced by the reverberations from the caves behind the pouring waters. At night-time he attuned to the wind's soothing sounds. He acquired healing powers. When the indigenous Tiwa people visited him to cure their babies' illnesses, they told him about the mountain's song. Nature holding counsel with her own, resetting a pattern of harmony, they said.

What is now simply called 'The Taos Hum' is a phenomenon of the infrasonic range. Its wavelength is between 17 and 342 metres (56 and 1,122 ft) in length and can propagate over large distances without being reflected or absorbed by obstacles. The Hum might be man-made or natural. The source of a sound with such qualities makes it hard to trace and thus conspiracy theories flourish: aliens, supersonic weapons and electromagnetic vibrations have been considered. Investigations by the nearby Los Alamos National Laboratory, University of New Mexico, the Air Force Research Laboratory (formerly Phillips Laboratory) and Sandia National Laboratories could not find an answer.

There is the Taos Hum, but there also is the Largs Hum in Scotland (rumoured to be caused by submarines), the Kokomo Hum in Indiana (caused by a nearby factory) and even a hum in southern Germany (the federal government looked into it and has not found answers). The World Hum Map Database Project lists possible sources. As the sound is not recordable, one explanation traces its origin to microwave sound, where the thermo-elastic wave of acoustic pressure is absorbed directly by the soft tissues in the head. Sensed here, a signal travels by bone conduction to the inner ear. A buzzing, ticking, hissing sound is heard by no one but the medium.

> Superficially, the world has become small and known. Poor little globe of earth, the tourists trot round you as easily as they trot round the Bois or round Central Park. There is no mystery left, we've been there, we've seen it, we know all about it. We've done the globe and the globe is done.

So wrote D. H. Lawrence, who made Taos his home before he died. Yet the Taos Hum remains elusive. Whether it is the mountain singing or society not fully understanding its machines, inventions and fabrications remains a modern mystery. The Taos Hum might be an old sound with a young history or a new sound with an old history. We do not know.

26

One Square Inch of Silence

Orfield Laboratories, Minneapolis, USA | Haleakalā Crater, Maui, USA | Hoh Rain Forest, Olympic National Park, USA | N 48.12885°, W 123.68234°, 303 feet above sea level

In a book full of strange, bizarre sounds, one chapter on where they are not. 'To hear, one must be silent.' One of the first lessons impressed upon the young mage Sparrowhawk near the beginning of Ursula K. Le Guin's novel *A Wizard of Earthsea*. His master: one who tends goats, walks the forests, and soothed the great earthquake, long ago. His name: Ogion the Silent, written with the rune of the Closed Mouth.

In the worlds of both fantasy and reality, it is easier to heed speakers than listeners, easier to enact than embody. So in the saga, Sparrowhawk leaves his master to answer the siren call of adventure. So in our modern lives, we hurry from thing to thing to fill our days with sound and movement. But how often do we take the time to sink into silence and stillness?

What if a bout of quietness can transform our bodies and minds?

Should you find yourself pondering these questions, book a tour to visit Orfield Laboratories in Minneapolis, Minnesota. Housed inside a nondescript concrete building is an acoustic anechoic chamber. It is a room lined with foam wedges designed to eat sound and absorb echoes. Even the floor is fitted with deadening material for achieving a 'free-field' condition, in which sound waves can travel out and never return.

The Orfield room can no longer make the claim of being the quietest place on Earth; its Guinness World Record of -13 dBA in 2013 was broken by Microsoft Building 87 in Redmond, Washington State, recorded at -20.6 dBA in 2015. But this matters little to the human ear, which cannot register anything below zero decibels.

It is a myth that people will go crazy if they spend more than 45 minutes inside the chamber without sound and light. President of the Laboratories Steven Orfield simply wanted media outlets to spend a substantial block of time within so they could describe how absolute silence feels. Experiences can include the heightened perception of existing sounds, like clothes rustling or joints creaking. Your voice may not sound like yours. The body's internal flows and beats may come to the fore. Sensations of weightlessness and disorientation can accompany such extreme sensory deprivation. Some find this ordeal claustrophobic, stifling, anxiety-inducing, while others find it tranquil, peaceful, restorative. The absence of sound can be a presence in and of itself.

Some actively seek utter silence outside in the natural world. Self-professed sound tracker Gordon Hempton found his negative decibel Mecca inside the Haleakalā Crater on Maui. The chaos of the external world cannot reach the interior of the dormant volcano, towering more than 3,000 metres (9,842 ft) above sea level. Here there is only wind and sky, quartz sand and still stars.

Those who have trekked into the crater in search of beauty, respite or wonder attribute an otherworldly quality to the surroundings. Pulitzer Prize-winning poet W. S. Merwin distinguishes between Haleakalā's stillness and silence. Stillness needs to be perceived by a listener. Silence is immanent to place. '[Silence] is there in every molecule, and it's there when we look up at the night sky, and much further than we can see.' In this way Merwin both addressed and sidestepped that popular koan: if a tree falls in the forest and no one is around to hear it, does it make a sound? The element that remains is the silence, even if tree and sound and listener have never been or have long since ceased to be.

In one such space, Hoh Rain Forest located in Olympic National Park, lies a small red stone. This, marked 'One Square Inch of Silence', is yet another venture of Hempton's, established on Earth Day in 2005 as a sanctuary against the dangers of man-made sound pollution. The World Health Organization estimates that environmental noise accounts for the loss of more than 1 million healthy life years in Western Europe alone. Sleep disturbance from traffic. Cognitive impairment in children. Increased hypertension and heart disease. Hempton's reasoning for creating 'One Square Inch': to invent a centre from which the appreciation of silence can ripple out and serve as an antidote against the proliferation of noise from our busy modern lives. 'If you defend a point, it's simple and clear to understand, and you can defend the whole area.'

One Square Inch is not truly silent. Hempton distinguishes between natural and human soundscapes, between the raindrops dripping off Sitka spruces and the Growler jets training nearby at the Whidbey Island Naval Air Station. But the mossy undergrowth and evergreen canopy do not discern which sounds they absorb. The ancient-growth forest itself serves as an anechoic chamber, but one that is never static, full of growth and decay, always breathing, making, changing, simultaneously living and dying.

Dwelling within the heart of the forest is the northern spotted owl, an endangered bird reliant on an endangered silence. The comb-like shapes of its wing feathers make it a noiseless flier; it chooses when and where it wants to be heard. The signal call: a series of four-noted hoots. Deep, pure tones forming a sentence that carries far on still nights. 'The word must be heard in silence.' One of the final thoughts imparted by the old mage Sparrowhawk to a young king towards the end of Ursula K. Le Guin's novel *The Farthest Shore*. 'There must be darkness to see the stars,' the Pacific Northwest author writes, inspired by her beloved forests, always close by.

27

Acoustics Accidental and Incidental

Grand Central Station, New York City, USA | St Paul's Cathedral, London, UK | The Amphitheatre of Epidaurus, Greece | Temple of Kukulcan, Mexico | Nijo Castle, Kyoto, Japan | Anyang, Gyeonggi, South Korea | Lancaster, California, USA | Metro Station Network, Washington, DC, USA

Of the 750,000 people who pass through Grand Central Terminal every day, a fraction will stop outside the Oyster Bar in the lower concourse to face one of the four corners at the domed intersection. Chances are that they are not merely admiring the herringbone tilework – made by renowned Spanish building engineer Rafael Guastavino – but are conducting a sound experiment with their compatriot standing in another corner across the walkway. This is the Whispering Arch, where amid the bustle and clamour, one can stop to have an intimate chat with both the building and someone listening 10 metres (33 ft) away. No need for string cups or noise-cancelling headphones – the low arches, tight-set tiles and vaulted ceiling all work together to carry the softest of murmurs clearly from one ear and one corner to another.

For another space renowned for this acoustic quirk, climb the 259 steps up to the stone gallery in St Paul's Cathedral in London. Designed by architect Sir Christopher Wren in the seventeenth century, the geometry of the central dome proved particularly conducive for conveying quiet sounds across the entire 33.7-metre (37 yd) diameter. The science behind this remained a mystery until 1878, when British physicist Lord Rayleigh named and described the whispering-gallery wave, noting that vibrations have a tendency to cling along a smooth concave surface and travel for long distances without scattering or dissipation.

Many buildings have since been discovered or constructed with whispering gallery qualities. The echo wall at the Temple of Heaven in Beijing, China, has reflected half a millennium's span of prayers for bountiful harvests. The Lover's Bench in Santiago de Compostela, Spain, held many a whisper between unmarried couples during the Franco years, when touching in public was prohibited. The wooden dome at the u.s. Capitol Building proved popular for both politicians and tourists until a gas explosion led to its less acoustically interesting (but fireproof) replacement in 1901. Through the confluence of design and physics, these sites have become accidental hosts to our spiritual pleas, our clandestine exchanges, our sweetest nothings.

Properties of sound arising from architecture have long been noticed and exploited across human history, even if construction did not take them into account and the mechanics were not fully understood. Built in the fourth century BC, the amphitheatre of Epidaurus in Greece was revered for its excellent acoustic properties, even if its architects did not realize the limestone seats formed a filter that dampened low-frequency crowd noises and reflected higher-frequency performer voices up to the back. To this day experts debate whether Mayan pyramid builders designed the Temple of Kukulcan so that clapping at the base will play back calls resembling a resplendent quetzal, a sacred local bird. Speaking of birds, the *uguisubari*, or nightingale floors, at Nijo Castle in Kyoto are so named due to the warbler-like chirps they emit when people walk on them. Legends speak of the groans and squeaks serving as an alarm system against thieves and assassins during the Edo period, while the sign on-site today states that 'the singing sound is not actually intentional, stemming rather from the movement of nails against clumps in the floor caused by wear and tear over the years.'

Sometimes acoustics might be intrinsic to the infrastructure itself. Melody roads are precisely grooved stretches of asphalt intended to sound out certain tunes when driven over at specific speeds. Five such roads were in operation in South Korea as of 2022 as part of an initiative by the Korean Highway Corp to reduce accident rates caused by tired or inattentive drivers. The first, located at Anyang, Gyeonggi, which has since been decommissioned, played a perfectly pleasant version of 'Mary Had a Little Lamb' at a speed of 100 kilometres per hour (62 mph). But even intentional designs can lead to accidental acoustics. Honda engineers built a melody road at Lancaster, California, as part of a publicity stunt. Due to a miscalculation in the groove intervals, their rendition of the iconic 'William Tell Overture' proved completely out of tune and borderline unlistenable. After the road was paved over due to noise complaints from nearby residents, it was rebuilt in another part of town, retaining the exact same flaw.

Perhaps whether a sound is intended or not does not matter, as long as it proves desirable. An interesting sound need not be understood to be enjoyed; a keen ear coupled with a receptive mind will suffice. Both might be necessary to appreciate the accidental music generated by ageing escalators at the Washington, DC, Metro Station Network. Local music critic Chris Richards began to discover beauty in the travails of these machines as they creaked away in disrepair. To him, the escalator at the west entrance at Petworth was 'all honk and grind – the clatter of a hundred bop quartets cooking from 5 a.m. to midnight', while the lifts at Benning Road 'drone like an Indian tambura while arbitrary notes squeak and blurt'. Will the 39-second ride up to Connecticut Avenue from Farragut North resemble late-period John Coltrane free jazz to your ears? Your mileage with this score of rust and wails may vary.

28

Sound Over Sight as Sense: The World According to Whales

The Ganges-Brahmaputra-Meghna river basin, India | Baffin Bay, West Greenland | The Ligurian Sea off the coast of Toulon, France

An ancient river herald with a surprisingly deep voice. A real-life unicorn producing sound beams unlike any other. The secret lives of deep-sea leviathans chanced upon by one deep-space telescope. Here are three fantastic beasts traversing their respective worlds with the power of sound.

Toothed whales join bats as the other major mammal group to have evolved echolocation, a biological form of sonar. Along stretches of the Ganges-Brahmaputra-Meghna river basin where visibility is measured in inches, the Ganges river dolphin navigates entirely by acoustic sense, lacking functional lenses in their eyes. To 'see', this *vahana*, or animal vehicle of the river goddess Ganga, broadcasts a series of short broadband clicks, resolving reflected echoes into pictures of its surroundings. Compared to its marine cousins navigating open waters, the call of this freshwater denizen ranges an octave lower. No need to evolve high-pitched, long-ranged systems for *Platanista gangetica*, sole survivor in the most ancient lineage of toothed whales – a flexible neck and rail-thin snout are enough for seizing fish and prawns along muddy river bottoms. Researchers suspect that its inner ear, unique in shape among all cetaceans past or present, serves to filter out the acoustic clutter of its environment, helping it negotiate around sand bars, mangrove roots and everything in between, almost always sight unseen.

Baffin Bay, West Greenland. Imagine: months of utter darkness in frigid seas north of the Arctic Circle. Imagine: the desperate challenge of finding food beneath and air above a thick layer of pack ice. An antipodal world to the murk of the Ganges, but yet another where sight reveals itself as a deficient sense. Welcome to the realm of the narwhal. This medium-sized sausage-shaped whale sends out a thousand clicks per second, receiving signals through the fatty pads on its lower jaw. Sound helps it scan the water column, locate the closest breathing hole, track the whereabouts of its next meal – likely a passing halibut or Arctic cod. Other creatures can echolocate, but none possess the intense focus of the narwhal. *Monodon monoceros* can broaden out its sonar transmissions to get a sense of its environment or tighten up its calls to render one target in high definition – think of an adjustable flashlight, but with a beam made of pure sound. Surprisingly the narwhal's most iconic trait, the male's 3-metre-long (10 ft) spiral tooth, plays no role in either navigation or overall survival. Females, which do not generally grow tusks, exhibit longer life expectancies.

～～～

The Ligurian Sea in the Mediterranean. The neutrino telescope ANTARES resides at a depth of 2,475 metres (8,120 ft). Constructed in 2008, it is designed to detect one of the most elusive particles in physics, but it has proven its worth not only in hearing the cosmos, but for tuning into oceanic depths. Two years of continuous sound data gathered through AMADEUS (ANTARES Modules for the Acoustic Detection Under the Sea) has revealed the year-round presence of sperm whales in the vicinity, likely drawn to the region's abundance of deep sea delectables. Real-time relaying of data has allowed for the hourly tracking of their distinct repertoire of calls: the usual clicks for general foraging at depth – directional, penetrating, ear-shattering. At 236 decibels, this largest of toothed whales is also the loudest animal in the world. Then there is 'the creak', a series of high-frequency clicks coming together to form the sound of a rusted door hinge. This is used when homing in on prey, almost exclusively 1-metre-long (3 ft) umbrella squids here in the Ligurian depths.

Sperm whale hunger is immense – each adult *Physeter macrocephalus* in a family pod may require up to a ton of food daily. Driven by need they drive their massive block-shaped heads and prune-skinned bodies down time and again, entering a realm hostile to all those who dwell above. Alone in this cold, crushing darkness, sound serves as the guiding star for sustenance and salvation. When the whales return to the surface, back to the warmth of sunlight and the lapping of waves, they will use sounds for another purpose – as codas to those beyond our ken, to speak with one another, to commune, as family.

29

Geophony, Biophony, Anthrophony: The Three Sound Types Beneath the Seven Seas

Challenger Deep seabed, southwest of Guam | Bay of Sainte Anne du Portzic, France | Leigh Marine Reserve, New Zealand | Sable Island, Canada | The Baltic Sea

Sound carries. So the saying goes and seems to ring true whenever a soprano's voice fills an opera house or the rumble of thunder echoes across a valley. But air in reality is a poor medium for sound, which travels quicker and farther in denser materials, like iron (15 times faster) or, better yet, diamond (35 times faster). While we do not live on a diamond planet like 55 Cancri e in the constellation of Cancer, we do inhabit a planet mostly covered with water, an excellent conductor in its own right and a host for a multitude of acoustic phenomena we have scarcely begun to fathom.

We know more about the surface of Mars than the Earth's oceans. So the saying goes and seems to ring true to this day; only one-fifth of the world's sea floors have been mapped by modern sonar. For many, the mysteries beneath the waves were first brought to colour and life by the films of French explorer Jacques Cousteau. But contrary to the title of his groundbreaking documentary *Le Monde du silence*, his beloved oceans in reality abound with sounds, many of which we are only just learning to tune into.

2015. Above the Challenger Deep seabed southwest of Guam, the National Oceanic and Atmospheric Administration, Oregon State University and the U.S. Coast Guard sink a titanium hydrophone 11 kilometres (7 mi.) down to listen in on the Mariana Trench. A metric ton of water bears down on every square centimetre of equipment at this depth; the process takes six hours to ensure that the instrument's ceramic housing does not crack under the pressure. To the surprise of researchers, the hydrophone picks up a constant stream of noise over the three weeks of recording. Chaos from a category 4 typhoon raging on the surface. Tremors from a magnitude 5 earthquake from a nearby ocean crust. Other strange signals recorded around the Mariana Trench include what has come to be known as the 'Western Pacific Biotwang'. Characterized by 'a complex structure, frequency sweep, and metallic conclusion', it was likely generated by a nearby population of dwarf minke whales.

'You would think the deepest part of the ocean would be one of the quietest places on Earth,' mused Robert Dziak, oceanographer and chief project scientist. But it seems that while light has forsaken this region of the ocean known as the hadal zone, sound has no trouble reaching and reverberating across this most inaccessible of environments.

The advent of underwater hydrophones has taught us that, like the hills, the seas are alive with the sound of music – just not with tunes tailored for our land-based ears, nor with singers or instruments we have come to expect. In the Bay of Sainte Anne du Portzic, France, spiny lobsters rub an antenna nub against a file under their eye – imagine dragging a bow over the strings of a violin – to produce defensive squeaks that can be heard up to 3 kilometres (2 mi.) away. In the Leigh Marine Reserve, New Zealand, grazing kina urchins scraping kelp off rocky reefs join in an evening chorus amplified by their skeletons serving as miniature Helmholtz resonators – think blowing air across the top of a jug. Across the Bahamas and in tropical waters around the world, the culprit behind a pervasive static that resembles frying bacon was revealed to be the snapping shrimp, which contract their claws so quickly that imploding gas bubbles are formed. The din generated by large groups of these 5-centimetre (2 in.) critters is so loud that it interfered with the detection of enemy submarines during the Second World War.

The field of soundscape ecology is in its infancy; these creatures comprise merely a few of the myriads of players in this aquatic orchestra. Far from being silent and tranquil spaces, we now know the oceans snap, crackle and pop with crustaceans; drum, grunt and croak with fish; click, chirp and wail with whales, all forming distinct, dynamic compositions that are not only heard, but felt and embodied by all that dwell in the depths.

Today human activity threatens to both fill and purge the seas of sound. Back at the bottom of the Mariana Trench, the drone of propeller noise can be heard as skyscraper-sized container ships pass through Guam to China and the Philippines. Off Sable Island near Nova Scotia, Canada, vessels are prospecting for deep-sea oil and gas for months on end using seismic airguns that fire underwater blasts as loud as 260 decibels. Sounds travelling across a liquid medium are not only heard, but felt and embodied; for some, as shock, as hearing loss, as a constant cacophony from which there is no escape or relief.

In the Baltic Sea, warming waters from climate change and nutrient run-offs from fish farms fuel the runaway growth of filamentous algae. The thick strands blanket eelgrass meadows on sandy seabeds, suffocate bladderwrack forests on rocky shores. As the filaments rot and decay, they drain their aquatic surroundings of oxygen, life, even sound. The resulting dead zones are muffled, utterly serene. The hydrophones pick up nothing.

30

'That's Not What It's Supposed to Sound Like': Bizarre Bird Calls from All Seven Continents

What would a book about sound be without birds? It would be too easy to delve deep into blackbirds singing in the dead of night or the ethereal melodies of meadowlarks. But this is an atlas of strange sounds, not a treatise on beautiful ones.

Trek across the world's continents and there will be birds on each, greeting you with bizarre calls. On the lonely, wind-scoured tip of the Antarctic Peninsula, gentoo penguins call to one another in what scientists term an 'ecstatic display', but can only be described as a clown horn honking. Across the coastal regions of Angola and much of southern Africa, grey go-away birds sound less avian and more petulant, filling the skies over acacia savannahs with the nasal whines of misbehaving children.

The strangeness of these vocals may stem from the fact that the calls do not seem to match the callers. The common loon, appearing so reserved on the reverse of the Canadian dollar, is not supposed to have a haunting wail and an unhinged laugh.

In contrast, the bald eagle comes across as anything but bold, with squawks resembling a high-pitched kettle instead of the trademark shriek Americans associate with their national bird (that actually comes from a red-tailed hawk). And one expecting gentle hoots while strolling through forests under a full moon might come away not only disappointed but terrified: barn owl calls in reality resemble hoarse, blood-curdling screams.

A bird's primary vocal organ, called a syrinx, is located at the base of the trachea. Yet many unique bird sounds do not make use of this unique feature at all. Both the Swinhoe's snipe and the pin-tailed snipe whirl across Russian wetlands with stiffened tail feathers in a behaviour known as 'drumming' – the sounds they produce range between a miniature jet taking off and a futuristic hover-car touching down. More mechanical sounds from feathered sources are in South and Central America, home to one-third of all bird species. The club-winged manakin of Colombia and Ecuador produces an electric-toothbrush hum by vibrating its feathers at a resonance frequency of exactly 1,498 hertz. The crested oropendola from Trinidad and Tobago calls with a signature similar to the chip tunes of a 1980s alien invasion video game. Then there is the white bell bird of the Guianas, whose yell resembles the testing signal from an emergency broadcast system. Male birds seem to direct their ear-splitting 125-decibel communiqués at prospective mates, at close range. 'When you watch them, it looks like they [the females] don't like it,' notes biologist Jeffrey Podos. 'It's pretty socially awkward.'

Then there are birds that can switch with ease between producing sounds we deem natural and those we think of as artificial. In southeastern Australia, the superb lyrebird is a master mimic, drawing equal inspiration from the laughing kookaburras of its wild habitat and the hammering of nails in its captive enclosure. Among the vast repertoire of pitch-perfect imitations by 'Chook', Adelaide Zoo's celebrity lyrebird over the course of his 32-year life: a power drill from the construction of a panda enclosure; the trill of triggered car alarms in the parking lot; the shutter-snap and motor drives from the cameras of overeager birders.

Some birds may even embrace the sonic chaos of the urban environment as a challenge, singing louder and with greater complexity in response to being in a noisy city. The thriving population of nightingales in Berlin has been recorded as singing over 2,300 types of songs. Researcher and violinist Charlotte Schneider described some of these verses as containing recurring motifs 'in the form of trills, rasping buzzes, hard beats in which tones and syllables seem to be strung together in an almost techno-like fashion'. These are not the same melodious plots that once inspired English romantic poets to odes that linger long past their frail lives. But beauty has always been in the eye (or in this case, the ear) of the beholder. We might do well to remember that most songs sung on Earth were never meant for us, that we are merely eavesdroppers on a grander symphony.

31

Sensing the Sound of a Landscape through Rock and Stone

Chapel of Sound, near Chengde, China

A barren, timeless landscape in the mountains of Transcaucasia. A young boy and his father witness an ancient ritual of the local singer-poets: 'Once every 20 years we Ashokhs meet here to test the power of our art. This valley is unique. Only a sound of special quality will make its stones vibrate. He who can produce this sound will be the winner.' One bard after another sings into the valley and on to the rocks. One of them hits a mystic frequency and the valley sings back. He is rewarded with a lamb. Filmmaker Peter Brook's opening scene of *Meetings with Remarkable Men* captures a memory of the mystic and composer Gurdjieff, whose lifelong quest for finding the connection between matter and sound may have originated in this valley.

Whispers of wind and raindrops animate a greener and lusher valley in another part of the world: Chengde, approximately two hours north of Beijing. An opulent but now empty palace bears witness to the summer retreat of the former emperor and his entourage. The Great Wall of China cuts through the scenery and divides it into two sides of the same place. Valleys, mountaintops, rivers and brooks, patches of grass, bushes, the home of birds. In the midst, as if thrown into it from above, a man-made boulder: the 'Chapel of Sound'.

In 2021, OPEN, a Chinese architectural firm, inserted this massive monolith into the valley. It is a Brutalist superstructure that appears heavy and ethereal at the same time. Fabricated from concrete mixed with an aggregate of local mineral-rich rocks, the 'Chapel of Sound' combines cutting-edge architecture, aesthetics and engineering with the sole purpose of enabling humans to listen to the landscape. Designed like a modern version of an amphitheatre, the space is intended to 'collect, reflect, and resonate nature itself'. The architects strove for sound itself to shape the building, furnishing it with an open roof to invite raindrops, winds and the ever-changing light to play along the structure's shell. The building process itself was a lesson in high-tech collaboration: computers simulated the flow of air and sound through every corner, edge and niche; thousands of small timber plates were carved and applied; and more than 10,000 pieces of rebar were bent and fixed together like a jigsaw puzzle. 'How can we be humble enough to hear what nature is murmuring to us?' the architects ask in their informational brochure.

In the Transcaucasian languages the word for a singer or bard, the ashik, stems from the root *ashiq*, 'to love', in Arabic. This most likely goes back even further in time and meaning to Avestan, a Zoroastrian language, where to love translates to more active notions of 'to wish, desire, and to seek'. The singers of Transcaucasia participated in competitions to prompt each other in such processes of seeking. When they weren't evoking the stones, they indulged in verbal duels, asking each other improvised riddles. One ashik would ask another, 'Tell me what falls to the ground from the sky?' And the answer would grow from a place within: 'Rain falls down to the ground from the sky.' After a pause he then might ask back: 'What remains dry in water?' And sweetly the answer was sung: 'Light does not become wet in water.'

How do we grow humble enough to hear what nature is murmuring to us? It is perhaps ironic to ask this question next to the Great Wall of China, a structure that has long withstood the erosion of time but is now crumbling under the exposure of acid rain. Slowly its piled-up bricks are returning to nature. What echoes will the fallen stones throw back in 10,000 years' time? Who will be there to listen to the sounds they cast?

32

The Body Fields and the Works of Jacob Kirkegaard

Copenhagen, Denmark | *Opus Mors* (record) | Body fields, Forensic Anthropology Center, Texas State University, San Marcos, USA

A night in Copenhagen, March 2022. The street lamps burn with a yellow tinge, sending just enough lumens to call it light. The wind sends chills over the cobblestones, a reminder of the sea nearby. A second-floor apartment, built some four hundred years ago, wooden floors creaking, doors closing with resistance. The breathing of a sleeping child in the room next door seeps into the atmosphere of the room where we sit. Sound artist Jacob Kirkegaard puts on a record. We lean back and listen.

Record 1: Opus Morturarium

First nothing, it seems. An emptiness, a bleakness, revealed in the absence of sound. Then, a low, deep kind of hiss, a constant parabola in a sphere of void, the background hum of a ventilation system. And behind, weak but present, the machinery that powers it. Sterile, clean and functional. 'In the morgue, bodies are lying underneath white sheets on these rolling tables. I placed the omnidirectional microphone in the room and left, because my physical presence would disturb the recording,' Kirkegaard explains in an interview with BOMB magazine. When we listen to it, he says nothing, but something seems to leak from the mechanics of the recording into the room. A loneliness, absolute.

Record 2: Opus Crematio

Metal machinery. A fire blasting. Burning flames. An audio rendering of heat, all-devouring. A moment of silence. Then some crushing of bones and the pouring of ashes. Heavy and powerful, engineered. Kirkegaard attached vibration sensors directly to the surface of the crematorium oven, translating the smallest stirrings into haunting sounds. The efficiency of leaving no traces of the living made manifest. Today, mobile crematoria are part of standard-issue war equipment.

Record 3: Opus Putesco

The body fields in Texas are a testing ground where donors give their bodies to science. Here, the forensic scientists learn how to deduce the time of death and its circumstance.

Like a puzzle, each emanation of sound triggers another mental picture. Insects humming and buzzing, all sorts of flies. A dry hot landscape, some shrivelled bushes. The mind yearns to piece these trembles into scenes, colours, crime. Red earth, a cloudless sky above. A meditation on a landscape filled with life. Then, maggots busy chewing, so much hunger, so much appetite. There is reassurance in this sound, a steady working.

AD 1247. In Hunan province Song Ci, a judge, tries to prove a murder case among the villagers. Every farmer is called to line up with his sickle in the central square. They stand in neat rows, holding up their tools, silver crescent and moon-shaped, polished and clean. But then the flies swarm towards one sickle. Only one. While the blood has been washed off, the insects are still drawn to its invisible traces. Song Ci would write the first entomological account of forensics, *Collected Cases of Injustice Rectified*, later to be translated for the world beyond.

What do insects tell us about the decomposing of a body? Like clockwork, each creature has their window to enter the scene. Blowflies, flesh flies, house flies, cheese flies, black soldier flies. Later the maggots, wasps, bees, mites even.

With a steady hand wrapped in plastic, Kirkegaard holds his microphone inside the carcass. 'That kind of positive conflict between all your associations or fears around death and then hearing something concrete, unfamiliar, even beautiful, in the decomposition of a corpse eases your mind, or certainly changes your relationship to that process. It is nature itself doing its job.'

Record 4: Opus Autopsia

A clean cut with a very sharp tool. The pulling away of skin. The scraping free from the skull, then a quivering plop. 'This is the brain,' Kirkegaard will say. We hear more cuts, flesh being pulled away, bones being broken. 'Listen to how different the heart sounds when put on the scale. It is a muscle, slightly jumpy.' And yes, the heart sounds very different to the brain. We find ourselves in the presence of a master craftsman, their every movement precise, an austere choreography.

While the sounds of dying might be deeply uncomfortable to the living, often accompanied by heavy breathing, moans of aching and fear, death itself is regarded as the eternal silence. But is it really? Kirkegaard set out to record the sounds surrounding and emanating from the dead human body. He focused on the intermediate time between the occurrence of death and the moment the corpse is put underground. *Opus Mors* is a record unlike any other and cannot be found on the Internet. One has to handle it, put it on, play it. Not for fun or curiosity. But as a meditation on life.

33

ASLSP:
As Slow As Possible

St Burchardi Church Halberstadt, Germany

Once there was an ocean here. Now the Kyffhäuser Mountains rise above the fertile plains of Germany's Saxony-Anhalt. Winds pick up around their tops, warm strands of air spiralling upwards. The wings of ravens flap in this breeze, drawing a pattern of flight, wide circles, keen eyes on the landscape. Every one hundred years, so the legend goes, King Barbarossa, sunken in slumber in the depths of these mountains, will wake up to check on the ravens. If ever they leave, a new era has arrived.

No wonder the nearby city of Halberstadt has a soft spot for eternity. Canned sausages were invented here in 1896, and it is home to a collection of 18,000 stuffed birds. The late Mrs Schraube's (1903–1980) former home, with its unchanged interior of 100 years' worth of bourgeois habitat, was turned into a museum. Don't let anything slip away from the past.

Hold on to life by preserving its remains. And then, in 2001, an organ was installed in the medieval St Burchardi church, playing ORGAN²/ASLSP by John Cage, or 'As Slow As Possible'.

How slow is as slow as possible? This question has become a philosophical and technical one, discussed at the Tage für Neue Orgelmusik in the Black Forest in 1998. Since an organ can theoretically be played as long as air is bellowed into its pipes, a piece of music can be as long as the lifespan of the instrument. Halberstadt is home of one of the older examples of a modern organ, built in 1361. When the conference took place, the organ was 639 years old. This then would have to be the length of the organ piece, the attendees decided, casting their gaze into the far future.

ORGAN²/ASLSP began with a pause that lasted seventeen months. Only the electronically operated wind machine under the bellows could be heard. For many a moon cycle, nothing but the long inhale and exhale, the hissing of the wind filling and leaving the lungs of the project. The first tones were played on 5 February 2003. Abruptly they switched on. One long-lasting G♯4, B4, G♯5, reverberating within the stone walls, filling the space with a hallowing presence, a sonic drone, sublime time made audible.

This performance has no player. Sandbags hold down the pedals for the pipes; the electric motor runs with an emergency back-up. At certain calculated dates new pipes will be added and new tones will sound. 'Anyone who is present for a sound change in the Burchardi church is confronted with the fleetingness of the moment,' writes Christoph Bossert, who calculated the sound changes until 2072.

At other times pipes will be taken away. Some time in the future another pause will linger; the silence it brings will need to be endured until the sound comes back again. The last tone in Halberstadt will ebb out in 2640.

One of the initiators of the project is Rainer O. Neugebauer, a professor of administrative sciences. Hence the almost remote approach to the project, which hinges on the project's organization in the distant future. To fund ORGAN^2/ASLSP, a sponsorship system was established. Finance one year of this long musical piece and your name will be engraved on a plaque. One year, one sponsor, one thousand Euros to keep a tone until the next one sets in. The first one hundred years were quickly sold, along with the years around the end. The in-between years remain voids that will need to be filled for the music to continue.

So far the inscriptions on the plaques tell us more about our age than the forthcoming one. 2085: We did it our way. 2186: To commemorate the nuclear disasters of Chernobyl and Fukushima. 2298: *Unendlich ist die Weite des Alls. Hellgrün das Gras auf den Gräbern* (The vastness of space is infinite. Pale green the grass on the graves). 2451: Yesterday was this morning.

'Ideally, it would do for thinking about time what the photographs of Earth from space have done for thinking about the environment. Such icons reframe the way people think.' So said Stewart Brand about projects executed by the Big Here and the Long Now Foundation. This could apply equally to projects like the Future Library project in Norway and this organ performance in Halberstadt.

To enter the medieval architecture of the St Burchardi church is to enter a space filled with sound. There will never be more than four tones playing at once, reverberating within the walls that once served as a pigsty and a distillery during the brief historic footnote that was the years of the German Democratic Republic. While some living nearby complain about the tinnitus-like noise, others profit from the culturally inclined tourists who would otherwise never choose Halberstadt as their destination. The town itself is a fraught territory: other forces have longed to reunite with eternity, like the local Nazi leagues, more than ready to celebrate their thousand-year Reich on a bigger scale than in their local pubs.

Cage wanted a music that was free of hierarchies. No tone should be more important than another. His contemporary Tōru Takemitsu, with whom he enjoyed collaborations and mushroom-picking trips: 'The naive and basic act of the human being, listening, has been forgotten. Music is something to be listened to, not explained. John Cage is trying to reconfirm the significance of this initial act.'

ASLSP might mean more or different things than as slow as possible. When Cage wrote the piece in 1985 – first for piano and two years later for organ – he provided contradictory performance notes: 'The title is an abbreviation of "as slow as possible". It also refers to "Soft morning city! Lsp!" the first exclamations in the last paragraph of *Finnegans Wake* by James Joyce.' It took Joyce seventeen years to write his obscure novel, which is filled with strange words, portmanteaus and expressions, drifting between states of dream and waking. Sometimes a made-up word, impossibly long and impossible to read out loud, will fill the page like a thunderclap: 'bababadalgharaghtakamminarronnkonnbronntonnerronntuonnthunntrovarrhounawnskawntoohoohoordenenthurnuk!' But it is the last paragraph that is seen as a recursion, linking the end with the beginning and forming a never-ending cycle. 'Soft morning city! Lsp! Lsp as but a gentle sigh, a mellow whispering breeze, the world fresh in its waking and the human body reacting to it.' Somewhere in those lines perhaps lies the meaning of the Halberstadt organ project: a massive nonsensical attempt to grasp a notion of eternity, not to preserve time but to keep it alive, a tender effort to let sound be sound and to be heard for what it is.

34

From Wax and Glass, Music and Voices: The Past, Present and Future of Sound Recording

The Mapleson cylinders, New York Public Library, New York City, USA | California Language Archive, University of California, Berkeley, USA | The Global Music Vault, Svalbard, Norway

Hit 'record'. To catch a passing sound and replay it later for listening is a technology not yet two centuries old, a flash in the annals of human history.

The year is 1860. Édouard-Léon Scott de Martinville is excited to have made a ten-second recording of the French folk-song 'Au clair de la lune' through his patented phonautograph. 'Au clair de la lune, mon ami Pierrot, prête moi . . .' Perhaps one day, he thinks, people will decipher the recording made on paper. Yet the idea of putting those signals back into the air for another's ear would occur neither to him nor anyone else until a decade and a half later.

The year is 1877. With his state-of-the-art foil-based phonograph, Thomas Edison shouts 'Mary Had a Little Lamb' and replays it to his head machinist John Kruesi. 'I was never so taken aback in my life,' Edison recalled five decades later at the golden jubilee of this world-changing invention. 'Everybody was astonished. I was always afraid of things that worked the first time.' With mainstream sound reproduction came the possibility for anyone to record and relive a slice of their sonic surroundings.

The year is 1900. Metropolitan Opera librarian Lionel Mapleson purchases an Edison phonograph, complete with the latest wax cylinders, to capture rehearsals and performances. After testing it out in the prompter's box and up in the flies, Mapleson places the machine with its oversized flared horn on a catwalk above the centre of the stage. For the next three years he captures an assortment of arias and scene excerpts, providing a cross-section from what many considered the golden age of opera. Today the Mapleson collection of 135 wax cylinders, housed at the New York Public Library, represents the oldest documentation of live musical performances from the first decade of the twentieth century. Artists at the opera house seem to sound more unfettered than on their commercial recordings, notes Edward Komara, Crane Librarian of Music at the State University of New York, singing with more swagger, sounding more heroic and larger than life – fitting for the operatic form.

The advent of wax cylinders allowed for the preservation of not only music but voices. Portable phonographs allowed ethnographers to go into the field and capture the languages of other cultures – many of which are no longer spoken.

California: once one of the most linguistically diverse regions in the western hemisphere. Today less than half of the ninety Indigenous languages still have living speakers. Beginning in 2015, linguist Andrew Garrett from the University of California, Berkeley, was part of a collaboration to digitize and restore audio recordings from 3,000 early twentieth-century cylinders held by the Phoebe A. Hearst Museum of Anthropology. No need for needles to read wax made brittle by time – a light-based optical scanning system developed by physicist and audio preservationist Carl Haber is able to convert the etched songs, rituals and stories into digital sound files for future research and revitalization efforts.

Due to cultural considerations, recordings are made available only upon request. Garrett describes the project as 'digital repatriation of cultural heritage to the people and communities where the knowledge was created in the first place'. Several recordings in Salinan from the Californian Language Archive are available to the general public. Their titles: 'My Trip to San Francisco', 'Fighting Forest Fires', 'Spanish-Salinan Brief Vocabulary I (Numerals 1 through 10)' and 'Spanish-Salinan Brief Vocabulary II'. The last speaker of Salinan died in 1958.

Everything will eventually succumb to entropy. Wax cylinders crumble and mould. Hard drives demagnetize and fail. Media archaeology posits that the formats we place our faith in are only slightly less ephemeral than the original transmission waves. At the Global Music Vault in Svalbard, Norway, an initiative to develop a longer-term solution. The objective of 'Project Silica': to encode mass amounts of music and other audio data on to glass plates impervious to flooding, extreme temperatures and electromagnetic radiation. The first proof-of-concept plate installed at the vault will include recordings from the International Library of African Music, concerts by Stevie Wonder and Manfred Mann, and works by multimedia musician Beatie Wolfe.

The glass format is expected to last 10,000 years, one whole span of recorded human history. Will anyone possess the equipment and knowledge base to play back our species's catalogue across deep time? In this we have no choice but to follow in the spirit of Édouard-Léon Scott de Martinville as he was perfecting his phonautograph – leave it up to the future to decipher our collective sonic legacy.

35

The Sound of Manipulation: Sonic Warfare and Propaganda

Tempelhofer Feld, Berlin, Germany | Telefunken mushroom loudspeaker, German Historical Museum, Berlin, Germany | Banks of the Rur River, Germany | Henry Ford Factory, Detroit, USA | Hau Nghia Province, Vietnam | Streets of Pittsburgh, USA

1 May 1937. Tempelhofer Feld, Berlin. An agitated, short man delivers a speech, spitting his words into cabbage-sized microphones. The choreographed gestures of his performance are visible only to the cameras and those standing in his immediate vicinity, but his words resound loud and clear to the far corners of the field – thanks to Telefunken's newly developed mushroom-shaped loudspeakers. Every syllable he speaks is broadcast to a sea of an estimated 1.5 million listeners.

Spring 1944. An odd entourage of designers, artists and illustrators arrives at the banks of the Rur River in Germany to support the Allies' forces. Equipped with inflatable fake rubber tanks and a recorder full of sound effects, they have come to fight without bullets. Months earlier at Fort Knox they recorded sounds associated with military manoeuvres using state-of-the-art-technology and can now play them back, adjusting new narratives according to conquerable territory: an approaching tank, thousands of soldiers' feet stamping in the mud, planes flying by. Up to 25 kilometres (15½ mi.) away, the amplified and DJ'd sounds of the so-called Ghost Army send shivers down the spines of their enemies.

Detroit, 1950. A Henry Ford car factory assembly line. The noise in this establishment is excruciating. Workers are said to show signs of collapse after only six months. The Muzak Stimulus Progression Program provides a backbeat to their fluorescent lives: a bit energizing in the morning, a little calming towards the afternoon, a choreography of energy levels to sync the workers with the machines. Singing is strictly forbidden.

In 1967 a boat floats on the murky waters of a Vietnamese jungle river. From it an eerie soundscape escapes into the dense and impenetrable bushes at the shore. First the clattering of coins, as practised by Buddhists during funeral rites, then the voice of a child crying in Vietnamese: 'Daddy, Daddy, come home with me.' More hollow and echoing sounds follow. Then: 'My friends. I come back to let you know I am dead.' 'Ghost Tape No. 10' (or 'The Wandering Soul') would also be played by the U.S. Army from loudspeakers on backpacks or from helicopters.

The power of sound. Invisible, penetrating and ferocious. It doesn't just tickle the fine hairs in the ear canal; it can mobilize a nation. The Nazis would use amplified voices as one of their main tools of mass manipulation, calculating that training speakers was less expensive than printing posters. Loudspeaker became a term of its own and gave rise to a new area of research. 'And so the National Socialist revolution unfolded: the Führer's breath swept across the heap of corpses at Weimar and brought the dry bones to life. One here, one there, a long row until, finally, the entire German people was rising from its grave, awoken by the voice of its prophet,' wrote Karl Kindt in *Der Führer als Redner* (The Leader as Speaker) in 1934. The spacing of loudspeakers at public rallies was cunningly crafted while the industry supported this demand with the development of new types of microphones and speaker systems put together in a factory in Groß-Ziethen, just outside Berlin. 'If only they didn't speak so loud all the time' was one of the few publicly voiced complaints against the oppressive Nazi ideology.

The soundscapes used by the U.S. Ghost Army during the Second World War did not kill or injure anyone while helping to pave the way to an Allied victory. The 1,100 men of this special forces unit would later rise to the creative elite in America: Arthur B. Singer illustrating fifty stamps of wildlife birds; Ralph Ingersoll funding *PM* magazine or Art Kane 'grabbing and twisting' the likes of Bob Dylan; Frank Zappa and the Rolling Stones for their iconic portraits. In 2022 President Biden signed the Ghost Army Congressional Gold Medal Act to recognize 'their unique and highly distinguished service in conducting deception operations in Europe during World War II'.

'Feelings, such as happiness and contentment, and even hearing rhythmic sounds, music, etc., are an aid toward increasing output,' wrote L. M. Gilbreth in *The Psychology of Management* in 1914. The Muzak Stimulus Progression Program eventually seeped out of the factories and into public spaces. Elevator music came to play at the White House, in outer space and in traffic terminals, numbing everyone with a cheerful tune. In 1967 the violinist Yehudi Menuhin had had enough of this constant sonic irrigation and delivered a passionate speech at the UN demanding the right to silence: 'We're overwhelmed today with background music. There's too much of it. It's kind of a background noise which absorbs too much of our consciousness.'

Operation Wandering Soul eventually had to be stopped as the Việt Cộng soldiers were not affected by the ghost tapes. One of the only reported cases where the sounds from the tapes actually led to their defection was when, instead of spectral sounds, the roaring of a tiger was delivered. The operation eventually had to be stopped as the Việt Cộng fighters started to target the source of the sounds, silencing loudspeakers and those carrying them.

24 September 2009. The *New York Times* reports on a civilian demonstration against the G20 conference in Pittsburgh:

> The police fired a sound cannon that emitted shrill beeps, causing demonstrators to cover their ears and back up, then threw tear gas canisters that released clouds of white smoke and stun grenades that exploded with sharp

flashes of light. City officials said they believed it was the first time the sound cannon had been used publicly.

'Other law enforcement agencies will be watching to see how it was used,' said Nate Harper, the Pittsburgh police bureau chief. 'It served its purpose well.'

The deployment of LRAD (long range acoustic devices) against the demonstrators in Pittsburgh counted as the first use of its kind in the United States. Karen Piper, a civilian who was present at the demonstrations, describes being exposed – suddenly and without warning – to a strong piercing sound from which she suffered permanent hearing loss.

Military research is continuously exploring the use of LRADs, claiming its effects are harmless while defending peace. Phone calls to leading science institutions would not be answered for reasons of classification. The use of LRAD by the military and police has been listed and ranges from aiming sonic waves at pirates attacking cruise ships, deterring immigrants from crossing European borders, and keeping demonstrators at bay, as at the Women's March against the inauguration of Donald Trump in Washington, DC, in 2017.

36

Humanity's Message for Whom? The Voyager Golden Record

167 astronomical units (AU) from the Sun, interstellar space

'In space, no one can hear you scream.' So states the tagline from the 1979 science-fiction horror film *Alien*. Neither our primal fear of the abyss nor our knowledge that sound cannot cross the vacuum of space has stopped us from trying to make ourselves heard out across the cosmos.

Launched in 1977, the Voyager 1 spacecraft is the most distant artificial object from Earth. On board the probe speeding 25 billion kilometres (15.6 billion mi.) away towards the constellation Camelopardalis is a phonograph record containing 115 images on one side and 90 minutes of audio on the other. It is an archive from the third planet orbiting an unremarkable yellow sequence star.

The Voyager Golden Record comes ready to play with a stylus at 16⅔ revolutions per minute. Included, a selection of music featuring but not limited to: Bach's Brandenburg

Concerto No. 2 in F major, Chuck Berry's 'Johnny B. Goode', 'Flowing Streams' performed by Kuan P'ing-hu. 'Here Comes the Sun' by The Beatles was naturally considered for inclusion; the copyright holder EMI demanded three times the project's budget in compensation.

Some titles for the 'Sounds of Earth' section: Wind, Rain, Surf. Birds, Hyena, Elephant. F-111 Flyby, Saturn V Lift-off. Footsteps, Heartbeat, Laughter. The last recording, scratchy and hard to make out, is reputed to be from Carl Sagan, project committee chair. This was confirmed by Ann Druyan, project creative director, Sagan's future wife and widow. 'She said that laugh was the very first impression of my dad,' their daughter Sasha would relay later. '[Mom] wanted to include [his laughter] . . . because she wanted it to live on forever.' Also on the record, sharing a title and track with pulsars, the brain waves of a woman madly in love. The EEG recording came from the 27-year-old Druyan, two days after deciding to get married to Sagan. Its sound has been described as 'a string of exploding firecrackers'.

Other tracks on Side B: spoken greetings in 55 languages arranged in alphabetical order, beginning with Akkadian, the lost Sumerian tongue, and ending with Wu, a modern Chinese dialect. One recording challenge involved not having enough female voices due to the scarcity of women delegates at the UN. Speakers of certain languages were unable to be located due to the project's short notice: Swahili was 'a regrettable omission'.

Woven into the background of human greetings – the songs of humpback whales, recorded by environmentalist

Roger Payne off the coast of Bermuda in 1970. Record co-producer Linda Salzman comments, 'During the entire Voyager project, all decisions were based on the assumption that there were two audiences for whom the message was being prepared – those of us who inhabit Earth and those who exist on the planets of distant stars.'

Transmitter and receiver. Speaker and hearer. Who is out there, listening? The location and trajectory of Voyager 1 can be accessed through the NASA Jet Propulsion Laboratory website. In AD 41977 the spacecraft will approach within 1.7 light years of Gliese 445, an unremarkable red dwarf star. Assuming there is someone listening, would they even wish to hear from us?

The 1986 science-fiction film *Star Trek IV: The Voyage Home* features the arrival of an alien probe on Earth. It broadcasts a message we do not understand. Humanity's weapons and technologies prove useless against it. The day is saved only by the reintroduction of a pair of humpback whales, acting as ambassadors on behalf of the planet. After a brief exchange, the probe leaves, seemingly satisfied. Our fictional heroes from this fictional future, much like our real scientists of the present, do not know what the whales said, only that their plaintive calls are more complex and haunting than we can imagine.

Will there come a day when whale song can save us from destruction? This seems unlikely. Science fiction remains fiction, and our salvation will most likely need to come from within. But hope is a mysterious quantity, sometimes manifested through the making of a message sent out into the void, regardless of the odds, with the noblest of intentions. 'It is, as much as the sounds of any baleen whale, a love song cast upon the vastness of the deep,' Sagan writes of the Golden Record project in his 1980 book *Cosmos*. Such sentiment might save and/or outlast us, in the end.

Bibliography

1 Sounds of the Universe: In Search of Harmony and Resonance

American Physics Society, 'Holmdel Horn Antenna', www.aps.org, accessed 15 August 2022

Brennan, P., 'Cosmic Milestone: NASA Confirms 5,000 Exoplanets', NASA, www.science.nasa.gov, 21 March 2022

Chandra X-Ray Observatory, 'Chandra "Hears" a Black Hole for the First Time', www.chandra.harvard.edu, accessed 9 January 2025

SYSTEM Sounds, www.system-sounds.com, accessed 15 August 2022

2 The Shape of Sound: The Singing Pillars of Hampi

Benka, Stephen G., 'Musical Pillars Made of Solid Granite', *Physics Today*, LXI/11 (2008), https://doi.org/10.1063/1.4796701

Keiter, Mark, 'Geologie Indiens und Hampis: Eine lange Geschichte in wenigen Worten', www.uni-muenster.de, accessed 20 December 2021

Modak, H. V. 'Musical Curiosities in the Temples of South India', *Impact of Science on Society 138/139*, UNESCO (1985)

Paes, Domingo, *A Forgotten Empire (Vijayanagar): A Contribution to the History of India*, www.gutenberg.org, accessed 20 December 2021

Raja, Sudra, 'Music from the Pillars', *The Hindu*, 4 June 2015

3 An Arms Race 50 Million Years in the Making: The Shadow War of Moths and Bats

Bat Conservation International, 'Bracken Cave's Batty Bunch', www.batcon.org, 27 August 2020

—, 'Mexican Free-Tailed Bat', www.batcon.org, accessed 18 August 2022

Corcoran, A. J., J. R. Barber, and W. E. Conner, 'Tiger Moth Jams Bat Sonar', *Science*, CCCXXV/5938 (2009), pp. 325–7

Corcoran, A. J., and W. E. Conner, 'Sonar Jamming in the Field: Effectiveness and Behavior of a Unique Prey Defense', *Journal of Experimental Biology*, CCXV/24 (2012), pp. 4278–87

Long, Tony, 'Feb. 26, 1935: Radar, the Invention that Saved Britain', *Wired*, 26 February 2008

Minkel, J. R., 'Bats Flew Before They Could Echolocate', *Scientific American*, 14 February 2008

Rydell, J., et al., 'Bat Selfies: Photographic Surveys of Flying Bats', *Mammalian Biology*, CII/3 (2022), pp. 793–809

4 Caves, Sounds and the Human Imagination

Amos, Jonathan, 'Indonesia: Archaeologists Find World's Oldest Animal Cave Painting', BBC News, www.bbc.co.uk, 14 January 2021

Fondazione Mida Musei Integrati dell'Ambiente, 'Grotte Di Pertosa-Auletta', www.fondazionemida.com, accessed 10 August 2022

Hoffman, Jascha, 'Q&A: Acoustic Archaeologist', *Nature*, DVI (13 February 2014), p. 158

Miyagawa, S., C. Lesure and V. A. Nóbrega, 'Cross-Modality Information Transfer: A Hypothesis about the Relationship among Prehistoric Cave Paintings, Symbolic Thinking, and the Emergence of Language', *Frontiers in Psychology*, IX (20 February 2018), https://doi.org/10.3389/fpsyg.2018.00115

Musée Archéologie Nationale, 'La Grotte Chauvet-Pont d'Arc', www.archeologie.culture.gouv.fr, accessed 10 August 2022

Reznikoff, I., 'On Primitive Elements of Musical Meaning', *JMM: The Journal of Music and Meaning*, III (Fall 2004/Winter 2005)

SoundZipper staff, 'Magical Acoustics of the Prehistoric Caves', www.soundzipper.com, 2 October 2017

Till, Rupert, 'Songs of the Caves', www.songsofthecaves.wordpress.com, 24 June 2014

UNESCO World Heritage Convention, 'Cueva de las Manos, Río Pinturas', https://whc.unesco.org, accessed 10 August 2022

5 Reconstruction of the Paleosonic: Bringing Back the Sounds of Past Life

Leach, Amy, *Things That Are* (Minneapolis, MN, 2012), p. 13

Museum of Modern Art, 'Lucy from the Series Back, Herebelow Formidable (The Rebirth of Prehistoric Creatures)', www.moma.org, accessed 8 August 2021

NPR Staff, 'Ever Wonder What A Woolly Mammoth Sounds Like?' All Things Considered podcast, www.npr.org, 23 July 2011

University of Helsinki, 'The Last Mammoths Died on a Remote Island', *ScienceDaily*, www.sciencedaily.com, 7 October 2019

6 Calling Out Across Far Distances: Sound Messages in Air and Water

Kolbert, E., 'The Strange and Secret Ways that Animals Perceive the World', *New Yorker*, 6 June 2022

'Landesrecht konsolidiert Vorarlberg: Gesamte Rechtsvorschrift für Verordnung der Landesregierung über Warn- und Alarmsignale, Fassung vom 11.11.2022', www.ris.bka.gv.at, 23 December 2024

Laufer, Berthold, 'Chinese Pigeon Whistles', *Scientific American*, 30 May 1908

Todd, B., 'Pigeon Whistles: The Closest Thing I've Ever Experienced to Heaven', *The Guardian*, 22 August 2014

Yong, E., *An Immense World* (London, 2022), pp. 231–4

7 Reverberations Across Time and Fate: The Oracle at Dodona

Acropolis Museum, 'Dodona: The Oracle of Sounds', www.theacropolismuseum.gr, accessed 10 October 2021

Anon., 'The Æolian Harp', *Scientific American Supplement*, 483 (1885), https://chestofbooks.com, accessed 17 August 2022

Greek Ministry of Education and Religious Affairs, Culture and Sports, 'Ancient Theatre of Dodona', http://odysseus.culture.gr, accessed 15 October 2021

Hawthorne, Nathaniel, *The Golden Fleece*, as told by Nathaniel Hawthorne (1853), www.argonauts-book.com, accessed 15 October 2021

Murray, Augustus Taber, trans., *The Iliad*, Chapter 16, www.americanliterature.com, accessed 15 October 2021

Nicol, D. M., 'The Oracle of Dodona', *Greece and Rome*, V/2 (1958), p. 128

8 Word on the Wind: The Curious History of the Aeolian Harp

Cavanaugh, C. A., 'The Aeolian Harp: Beauty and Unity in the Poetry and Prose of Ralph Waldo Emerson', *Rocky Mountain Review*, LVI/1 (2002), p. 25

Ebrahimji, Alisha, 'Here's Why the Golden Gate Bridge Sings in San Francisco Now', *CNN*, https://edition.cnn.com, 6 June 2020

Engel, Carl, 'The Project Gutenberg eBook, Musical Instruments', www.gutenberg.org, accessed 17 August 2022

Matteson, Robert S., 'Emerson and the Æolian Harp', *South Central Bulletin*, XXIII/4 (1963), pp. 4–9

Nelson, L., et al., 'Aeolian Harps and the Romantics', www.sites.udel.edu, accessed 17 August 2022

White, Chris, 'Eerie Instruments Played by the Wind', Atlas Obscura, www.atlasobscura.com, 22 August 2014

9 To Be Defined By Sound: Regarding the Cicada

Deutsch, James, 'Cicada Folklore, or Why We Don't Mind Billions of Burrowing Bugs at Once', *Folklife Magazine*, https://folklife.si.edu, 17 May 2021
Various artists, 'Utom: Summoning the Spirit', *Smithsonian Folkway Recordings* (1997), www.folkways.si.edu, accessed 19 February 2022
Warren, Mobi, 'Cicadas Still Sing', *Orion*, 2 June 2021
Wired 'Bug Expert Explains Why Cicadas Are So Loud', *Currents* podcast, episode 97, *Wired*, 24 May 2021
Zarrelli, Natalie, '"O, Shrill-Voiced Insect": The Cicada Poems of Ancient Greece', Atlas Obscura, www.atlasobscura.com, 23 June 2016

10 *Gagaku*: A Music that Crosses Space and Time

Ishikawa Sho, collected notes from profession as lecturer at Tokyo University, private.
Narabe, Kazumi, 'The Japanese Sho Mouth Organ of Mayumi Miyata – Giving Voice to the Natural World', The Japan Foundation, https://performingarts.jpf.go.jp, accessed 29 January 2022
Private interviews with Ko Ishikawa
Shimizu, Chatori, 'A Study on Japanese Instrumental Training Methods for Non-Japanese Students: Cases of Columbia University Mentor/Protégé Program (Gagaku)', *Senzoku Renso*, www.chatorishimizu.com/ronsou47, accessed 29 January 2025
Takemitsu, Toru, *Confronting Silence* (Berkeley, CA, 1995), pp. 3–12
Whatley, Catherine, 'Sho Player Ko Ishikawa Pushes the Boundaries of Gagaku with Free Improvisation', *Japan Times*, 2 January 2019

11 Sounds of the Cryosphere

Exeter Buch, electronic text, Exeter, Cathedral Chapter Library, MS 3501.
Kraus, Alexander, 'Der Klang des Nordpolarmeers', in *Weltmeere: Wissen und Wahrnehmung im langen 19. Jahrhundert. Vandenhoeck and Ruprecht*, ed. Alexander Kraus and Martina Winkler (Bristol, CT, 2014), pp. 127–49
Lee, Jane J., 'Watch and Listen to the Surprisingly Noisy Death of an Iceberg', *National Geographic*, www.nationalgeographic.com, 17 July 2013

Lopez, Barry, *Arctic Dreams* (London, 2014)
Magnason, Andri Snaer, *On Time and Water* (London, 2020), pp. 79–108

12 The Stories and Secrets of Singing Sands

China Discovery staff, 'Mingsha Mountain Crescent Lake Nature Park', *China Discovery*, www.chinadiscovery.com, accessed 16 June 2022

Dagois-Bohy, S. Courrech du Pont and S. Douady, 'Singing-Sand Avalanches without Dunes', *Geophysical Research Letters*, XXXIX/20 (2012), https://doi.org/10.1029/2012GL052540

Darwin, Charles, 'Chapter XVI: Northern Chile and Peru', in *The Voyage of the Beagle*, p. 2, https://darwin.thefreelibrary.com, accessed 16 June 2022

Fischer, Shannon, 'Singing Sand Dunes Explained', *National Geographic*, 31 October 2012

Jameson, Robert, 'On a Peculiar Noise Heard at Nakuh, on Mount Sinai', *Edinburgh New Philosophical Journal*, VIII (January–April 1830), pp. 74–5

Ouellette, Jennifer, 'Of Granular Material and Singing Sands', *Scientific American*, 17 November 2011

Polo, M., W. Marsden and J. Masefield, *The Travels of Marco Polo the Venetian*, www.archive.org, accessed 16 June 2022

Vriend, N. M., M. L. Hunt and R. W. Clayton, 'Linear and Nonlinear Wave Propagation in Booming Sand Dunes', *Physics of Fluids*, XXVII/10 (2015), https://doi.org/10.1063/1.4931971

13 The Humility Pipe

Bauer, Karlheinz, 'Die Orgelbauerfamilie Allgeyer in Hofen und Wasseralfingen', in *Aaleneer Jahrbuch* (Stuttgart, 1986)

Hesse, Hermann, *The Glass Bead Game* (New York, 1949)

Interviews with Cembalo musician Gösta Funck on keyboard instruments and phone conversation with Klais Orgelbau

Klais Orgelbau, 'Sprache, Stil und Landscaft', https://klais.de/m.php?pid=15, accessed 1 December 2024

Krahé, Detlev, et al., *Feasibility study on the Effects of Infrasound, Study Commissioned by the Federal Environment Agency, Dessau* (Roßlau, 2014)

Molkow, Wolfgang, 'Macht oder Andacht – Die Orgel im Film', *Neue Musikzeitung*, April 2021

Vernes, Jules, *20000 Meilen unter dem Meer* (Frankfurt, 2008)

14 The Untameable Ritual of Keening

Bourke, Angela, 'The Irish Traditional Lament and the Grieving Process', *Women's Studies International Forum*, XI/4 (1988), pp. 287–91

Lecossois, Hélène, *Performance, Modernity and the Plays of J. M. Synge* (Cambridge, 2022)

Naimon, David, *Between the Covers* podcast, interview with Doireann Ní Ghríofa, https://tinhouse.com, accessed 10 March 2025

Ní Ghríofa, Doireann, *A Ghost in the Throat* (Dublin, 2020)

Ní Uallacháin, 'Pádraigín: An tAmhrán Geal (The Song of Light)', www.youtube.com, accessed 1 March 2024

15 Death of a Sound and Age: Pining for the Foghorn

Foghorn Requiem, www.foghornrequiem.org, accessed 6 August 2022

Macdougall, Jane, 'An Ode to the Foghorn', *National Post*, https://nationalpost.com, 26 January 2013

MacKinnon, Charles, 'Foulis, Robert', *Dictionary of Canadian Biography*, IX (2003), www.biographi.ca, accessed 6 August 2022

Smellie, Sarah, 'Foghorns Cherished Part of Atlantic Canada Soundscape, But Do They Serve a Purpose?', *Canadian Press*, https://halifax.citynews.ca, 15 December 2021

Sweeney, Éamon, 'Foghorns Are Embedded in People's Minds', *Irish Times*, www.irishtimes.com, 12 May 2021

16 A Voice, an Echo, Silence

Brecht, Bertold, *An die Nachgeborenen* (Frankfurt, 1999)

Calvino, Italo, *Invisible Cities* (London, 1997)

Fontane, Theodor, *Der Große und Kleine Tornow-See*, in *Wanderungen durch die Mark Brandenburg in 8 Bänden*, vol. II: *Oderland*, ed. Gotthard Erler and Rudolf Mingau (Berlin, 1997)

Shikibu, Murasaki, *The Tale of Genji* (London, 2006)

17 Sounds of the Atomic Age

Agence France-Press, 'Bikini Atoll Nuclear Test: 60 Years Later and Islands Still Unliveable', *The Guardian*, 2 March 2014

Alexis-Martin, Becky, 'The Atomic History of Kiritimati – A Tiny Island Where Humanity Realized Its Most Lethal Potential', *The Conversation*, https://theconversation.com, 4 July 2019

Atomic Heritage Foundation, 'Castle Bravo', www.atomicheritage.org, 1 March 2017

BBC News, 'French Nuclear Tests Contaminated 110,000 in Pacific, Says Study', www.bbc.co.uk/news, 9 March 2021

Farrell, Thomas F., 'Groves and Farrell Watching Trinity', Atomic Heritage Foundation, www.atomicheritage.org, accessed 1 March 2022

Hersey, John, 'Hiroshima', *New Yorker*, 23 August 1946

LaFrance, Adrienne, 'The Sound of an Atomic Bomb', *The Atlantic*, 8 August 2017

Moruora Files: Investigation into French Nuclear Tests in the Pacific, https://moruroa-files.org, accessed 1 March 2022

Mott, Robert L., *Sound Effects: Radio, TV, and Film* (Boston, MA, 1990)

The Nuclear Weapon Archive, 'Operation Upshot-Knothole', https://nuclearweaponarchive.org, accessed 1 March 2022

18 Sounds from the Deepest Artificial Point on Earth

Bennett, Justin, 'Vilgiskoddeoayvinyarvi: Wolf Lake on the Mountains', https://soundcloud.com, accessed 10 January 2025

Carnegie Museum of Natural History, 'Stratavator – Benedum Hall of Geology. Carnegie Museum of Natural History. Pittsburgh, USA', www.youtube.com, accessed 10 January 2025

Kringel, Danny, 'Hilfe, wir haben die Hölle angebohrt', *Spiegel*, 26 March 2011

Nietzsche, Friedrich, *Also sprach Zarathustra*, vol. IV (Leipzig, 1891)

19 The Humming Fields and Meadows of the Altai

Haskell, David, 'Listening and the Crisis of Inattention – An Interview with David G. Haskell', *Emergence Magazine*, https://emergencemagazine.org, 21 April 2022

Vieser, Michaela, 'Suche nach dem verheißenen Land', *Zeitfragen Deutsch-landfunk Kultur*, www.deutschlandfunkkultur.de, 7 December 2016

20 Otherworldly Ordinary: The Found Sounds of the Fantastic

Burlingame, Jon, 'Hans Zimmer on "Dune" Score's Electronic Textures and Made-Up Choral Language', *Variety*, https://variety.com, 21 October 2021

Cole, Adam, and Ryan Kellman, 'Why Does a Frozen Lake Sound Like a Star Wars Blaster?' All Things Considered podcast, www.npr.org, 21 December 2016

Colorado Parks and Wildlife [@COParksWildlife], 'Ice at Steamboat Lake state park', X, www.x.com, 2 December 2021

Elgueta, Adriana, 'Mysterious Sounds Heard from Frozen Lake Sparks Bizarre Alien Theory', *The Sun*, 6 December 2021

King, Darryn, 'How Hans Zimmer Conjured the Otherworldly Sounds of "Dune"', *New York Times*, 22 October 2021

Kolganov, Alexey, [@alex.kolgan], Video of Lake Baikal, Instagram, www.instagram.com, 7 February 2021

Mandelbaum, Ryan F., 'Pew-Pew: What Do Lasers Actually Sound Like?' *Gizmodo*, 3 July 2019

Rinzler, J. W., *The Sounds of Star Wars* (San Francisco, CA, 2010), p. 54

Stearn, Maddie, 'What Was That? The Top 13 Foley SFX from Everyday Household Objects', www.storyblocks.com, 23 January 2017

21 Pay Attention: On the Sounds of the In-Between

Corbett, Sara, 'Do You Hear What I Hear?' *This American Life* podcast, episode 516, 17 January 2014

Eno, Brian, *Ambient 1: Music for Airports*, liner notes from the initial American release, PVC 7908 (AMB 001), http://music.hyperreal.org, accessed 10 August 2022

Great Big Story [@GreatBigStory], 'In Tokyo, These Trains Jingle All the Way', www.youtube.com, 3 August 2018

Harris, Jensen, [@JensenHarris] 'Why I Killed the Microsoft Startup Sound', www.youtube.com, 29 May 2021

Lysaker, John T., 'Brian Eno's *Music for Airports* 40 Years Later', OUPblog, https://blog.oup.com, 20 December 2018

O'Connor, Neil, 'Diffusing the Norm – Brian Eno's Music For Airports', *Divergence Press*, 14 February 2019

Selvin, Joel, 'Q and A with Brian Eno', *SFGate Chronicle*, 2 June 1996

22 The Not-So-Secret Love Lives of Fish

Clinton, Larry, 'The Humming Toadfish Festival', Sausalito Historical Society, 27 March 2019

De la Peña, Nonny, 'What's Making that Noise? Surprisingly, It May Be Fish', *New York Times*, 9 April 2008

Erisman, Brad E., and Timothy J. Rowell, 'A Sound Worth Saving: Acoustic Characteristics of a Massive Fish Spawning Aggregation', *Biology Letters*, XVIII/12 (2017), https://doi.org/10.1098/rsbl.2017.0656

Staugler, Betty, 'Black Drum', University of Florida Blogs, https://blogs.ifas.ufl.edu, 18 July 2016

23 Voices Within and Without

Beavan, V., J. Read and C. Cartwright, 'The Prevalence of Voice-Hearers in the General Population: A Literature Review', *Journal of Mental Health*, XX (2011), pp. 281–92

CYMRU, Hearing Voices Network, Wales, http://hearingvoicescymru.org, accessed 29 June 2021

Worthen, Molly, 'When God Talks Back', *New York Times*, 27 April 2012

24 Listening to Traces

Brandt, Jan, 'Die Leute wollen etwas Kuscheliges', *TAZ*, 23 October 1999

Burke, Claire, 'The Christmas Story of One Tube Station's "Mind the Gap" Voice', *The Guardian*, www.theguardian.com, 25 December 2019

Jarke, Philippa, 'Schauspielerin und Sprecherin', www.rbb-online.de, accessed 30 September 2022

Jones, Maggie, 'How Talking to the Dead Dislodged Some of My Sorrow', *New York Times*, www.nytimes.com, 6 July 2022

Messina Imai, Laura, 'How Japan's Wind Phone Became a Bridge Between Life and Death', https://lithub.com, 17 March 2021

Ní Ghríofa, Doireann, interview with David Naimon, 'Between the Covers Doireann Ní Ghríofa Interview', https://tinhouse.com, March 2022

25 Explaining the Inexplicable: The Taos Hum

Barker, Elsa, and Omar Barker, 'Hermit of the Mountain', *New Mexico Quarterly*, XXXI/4 (1961), pp. 349–55

Cowan, James P., 'The Results of Hum Studies in the United States', in *Community: 9th International Congress on Noise as a Public Health Problem (ICBEN)* (Foxwoods, CT, 2008)

Deming, David, 'The Hum: An Anomalous Sound Heard Around the World', *Journal of Scientific Exploration*, XVIII/4 (2004), pp. 571–95

Mühlhans Jörg H., 'Low Frequency and Infrasound: A Critical Review of the Myths, Misbeliefs and their Relevance to Music Perception Research', *Musicae Scientiae*, XXI/3 (2017), pp. 267–86

26 One Square Inch of Silence

Hanstock, Bill, 'The Bizarre Experience of Sitting Inside the "Quietest Room on Earth"', https://uproxx.com, 10 March 2017

Hempton, Gordon, 'Welcome to One Square Inch',
  www.onesquareinch.org, accessed 1 March 2025
Kahn, Brian, 'The Quietest Place in America Is Becoming a Warzone',
  https://gizmodo.com, 30 July 2018
Le Guin, Ursula K., *A Wizard of Earthsea* (Berkeley, CA, 1968)
—, *The Farthest Shore* (New York, 1972)
Swatman, Rachel, 'Microsoft Lab Sets New Record for the World's
  Quietest Place', *Guinness World Records*, www.guinnessworldrecords.
  com, 2 October 2015
Vendetti, Tom, 'Quietest Place on Earth', www.vimeo.com, 25 March
  2020
World Health Organization, 'Noise', www.who.int, accessed 1 October
  2024
Zellner, Xander, 'How Do Barn Owls Fly So Silently?', www.audubon.
  org, 13 May 2016

27 Acoustics Accidental and Incidental

Ball, Philip, 'Mystery of "Chirping" Pyramid Decoded', *Nature*,
  14 December 2004
Cho, Joohee, and Rebecca Lee, 'Singing Streets and Melody Roads',
  ABC News, 29 November 2007
Cox, Trevor, 'The Acoustics of Eavesdropping', *Slate*, https://slate.com,
  6 June 2014
Debczak, Michele, '7 Whispering Galleries from Around the World
  You Can Visit', www.mentalfloss.com, 10 March 2017
FitzGerald, James, 'How Does the Whispering Gallery at St Paul's
  Actually Work?' https://londonist.com, updated 13 January
  2023
Georgia Institute of Technology, 'Ancient Greek Amphitheater: Why
  You Can Hear from Back Row', *ScienceDaily*, www.sciencedaily.com,
  6 April 2007
Goldfarb, Kara, 'How Grand Central Terminal's Whispering
  Gallery Allows You To Hear Soft Words 30 Feet Away',
  https://allthatsinteresting.com, 8 June 2018
Patoway, Kaushik, 'Nightingale Floor: An Ancient Japanese Intruder
  Detection System', *Amusing Planet*, www.amusingplanet.com,
  27 April 2018
Rayleigh, Lord, 'CXII: The Problem of the Whispering Gallery',
  *London, Edinburgh, and Dublin Philosophical Magazine and
  Journal of Science*, XX/120 (1910), pp. 1001–4
Richards, Chris, 'Move Along with the Soundtrack of Metro's
  Screechy, Wailing Escalators', *Washington Post*, 14 January 2011

Simmons-Duffin, David, 'Honda Needs a Tune-Up', http://davidsd.org, 23 December 2008

28 Sound Over Sight as Sense: The World According to Whales

André, M., et al., 'Sperm Whale Long-Range Echolocation Sounds Revealed by ANTARES, a Deep-Sea Neutrino Telescope', *Scientific Reports*, VII/45517 (2017), https://doi:10.1038/srep45517
Biodiversity of India, 'Goddess Ganga and the Gangetic Dolphin', www.biodiversityofindia.org, accessed 17 May 2022
Jensen, F. H., et al., 'Clicking in Shallow Rivers: Short-Range Echo-Location of Irrawaddy and Ganges River Dolphins in a Shallow, Acoustically Complex Habitat', *PLOS One*, VIII/4 (April 2013), https://doi.org/10.1371/journal.pone.0059284
Klein, Joanna, 'Narwhals, Tusked Whales of the Arctic, See with Sound. Really Well.', *New York Times*, www.nytimes.com, 9 November 2016
Koblitz, Jens C., et al., 'Highly Directional Sonar Beam of Narwhals (*Monodon monoceros*) Measured with a Vertical 16 Hydrophone Array', *PLOS One*, XI/11 (November 2016), https://.doi.org/10.1371/journal.pone.0162069
University of California – Berkeley, press release, 'Whales Evolved Biosonar to Chase Squid Into the Deep', *Phys.org*, 5 September 2007

29 Geophony, Biophony, Anthrophony: The Three Sound Types Beneath the Seven Seas

Dziak, Bob, and Haru Matsumoto, 'Mariana Trench: Seven Miles Deep, the Ocean is Still a Noisy Place', Oregon State University Newsroom, https://news.oregonstate.edu, 2 March 2016
Gannon, Megan, 'Mysterious Metallic Sound in the Mariana Trench Finally Identified', *LiveScience*, www.livescience.com, 16 December 2016
Georgia State University, 'Speed of Sound in Various Bulk Media', http://hyperphysics.phy-ast.gsu.edu, accessed 18 August 2022
Jézéquel, Y., L. Chauvaud and J. Bonnel, 'Spiny Lobster Sounds Can Be Detectable over Kilometres Underwater', *Scientific Reports*, X/7943 (2020), https://doi:10.1038/s41598-020-64830-7
Marine Finland.fi, 'Eutrophication in the Baltic Sea', https://marinefinland.fi, accessed 18 August 2022
Moskowitz, Clara, 'Super-Earth Planet Likely Made of Diamond', www.space.com, 11 October 2012

NOAA Ocean Exploration, 'How Much of the Ocean Has Been Explored?' National Oceanic and Atmospheric Administration, ww.oceanexplorer.noaa.gov, accessed 22 August 2022

North Carolina State University, 'Oh, Snap! What Snapping Shrimp Sound Patterns May Tell Us about Reef Ecosystems', *ScienceDaily*, www.sciencedaily.com, 14 January 2016

University of Rhode Island, 'Sea Urchin Sounds (*Evechinus chloroticus*)', www.dosits.org, accessed 18 August 2022

—, 'Seismic Airguns', www.dosits.org, accessed 18 August 2022

Williams, Casey, 'The Bottom of the Ocean Is Surprisingly Noisy', *HuffPost*, www.huffingtonpost.co.uk, 3 March 2016

Woods Hole Oceanographic Institution, 'As Oceans Warm, Snapping Shrimp Sound a Warning', www.whoi.edu, 18 August 2022

30 'That's Not What It's Supposed to Sound Like': Bizarre Bird Calls from All Seven Continents

ABC News Australia, 'Loved Lyrebird Dies', www.abc.net.au, 28 December 2011

American Ornithological Society, 'A First Look at Geographic Variation in Gentoo Penguin Calls', *Phys.org*, 27 September 2017

Chung, Emily, 'The White Bellbird Is the Loudest Bird in the World', *CBC News*, www.cbc.ca, 22 October 2019

Keim, Brandon, 'Fastest Wings on Earth Show Extremes of Sexual Selection', *Wired*, 9 February 2010

Oltermann, Philip, 'Why Nightingales Are Snubbing Berkeley Square for the Tiergarten', *The Guardian*, www.theguardian.com, 13 April 2019

Romberg, Johanna, 'Her Singing Has Something of Techno', *Riffreporter*, www.riffreporter.de, 3 April 2019

31 Sensing the Sound of a Landscape through Rock and Stone

A Reimagination and Recreation of an Instrument of Sound, www.arup.com/projects, accessed 16 February 2022

Brook, Peter, and Jeanne de Salzmann, 'Meetings with Remarkable Men (1979): Full Transcript', https://subslikescript.com, accessed 21 January 2025

32 The Body Fields and the Works of Jacob Kirkegaard

Martin, Julie, 'Listening to the Heart: Jacob Kirkegaard', *Bomb Magazine*, 5 March 2021

Bibliography

Interview by the authors with Jacob Kirkegaard

33 ASLSP: As Slow As Possible

Joyce, James, *Finnegans Wake* (London, 1939)
Neugebauer, Rainer O., interview with Warren Senders for Music 4 Climate, at the 2021 United Nations (UN) Climate Change Conference (COP 26) in Glasgow
Takemitsu, Tōru, *Confronting Silence* (Berkeley, CA, 1995)

34 From Wax and Glass, Music and Voices: The Past, Present and Future of Sound Recording

Boutwell, Jane, 'The Mapleson Cylinders', *New Yorker*, 1 December 1985
Brice, Anne, 'For Native American Student, Reclaiming his Culture Began at Berkeley', *Berkeley News*, 5 June 2018
California Language Archive, 'The J. Alden Mason Collection of Salinan Sound Recordings', https://cla.berkeley.edu, accessed 9 August 2022
Dorrier, Jason, 'Microsoft to Archive Music on Futuristic Slivers of Glass That Will Live 10,000 Years', Singularity Hub, https://singularityhub.com, 12 June 2022
Edison, Thomas A., 'The Phonograph and Its Future', *North American Review*, CXXVI/262 (1878), pp. 527–36
First Sounds, 'The Phonautograms of Édouard-Léon Scott de Martinville', www.firstsounds.org, accessed 9 August 2022
Garrett, Andrew, 'Archives: Survey of California and Other Indian Languages', https://linguistics.berkeley.edu, accessed 9 August 2022
Global Music Vault, www.globalmusicvault.com, accessed 9 August 2022
Komara, Edward, 'Lionel Mapleson Cylinder Recordings of the Metropolitan Opera (1900–1903)', Library of Congress, www.loc.gov, accessed 9 August 2022
National Park Service, 'Origins of Sound Recording: Edison's Path to the Phonograph', www.nps.gov, 17 July 2017
National Public Radio, 'Reconsidering Earliest-Known Recording', *All Things Considered* podcast, www.npr.org, 1 June 2009
National Science Foundation, 'Linguistic and Ethnographic Sound Recordings from Early Twentieth-Century California: Optical Scanning, Digitization, and Access', www.nsf.gov, accessed 9 August 2022

PDR, 'Edison Reading Mary Had a Little Lamb (1927)', *Public Domain Review*, www.publicdomainreview.org, accessed 9 August 2022
Rosen, Rebecca J., 'Scientists Recover the Sounds of 19th-Century Music and Laughter from the Oldest Playable American Recording', *The Atlantic*, 26 October 2012

35 The Sound of Manipulation: Sonic Warfare and Propaganda

ACLU, 'Piper v. City of Pittsburgh', www.aclupa.org, accessed 7 November 2022
Eppinger-Jäger, C. 'Hitler's Voice: The Loudspeaker under National Socialism', *Intermediality*, 17 (2011), pp. 83–104
Gilbreth, L., *The Psychology of Management* (New York, 1914)
Goodman, S., *Sonic Warfare, Sound Effect and the Ecology of Fear* (Cambridge, MA, 2012)
Kindt, K., *Der Führer als Redner* (Hamburg, 1934)
'Neue Akustik-Waffe: Beschallen statt beschießen. Fraunhofer-Wissen-schaftler haben im Auftrag der Bundeswehr eine Schallwaffe entwickelt. Mit Lautstärke soll sie Soldaten die Gegner vom Hals halten', *Handelsblatt*, 17 May 2011
Quigley, M., 'Yehudi Raps', *Georgia Straight*, 1970, www.mjq.net, accessed 8 November 2022
'The Wandering Soul', *Smithsonian Magazine*, www.smithsonianmag.com, accessed 7 November 2022
Urbina, I., 'Protesters Are Met by Tear Gas at G-20 Conference', *New York Times*, www.nytimes.com, 24 September 2009
Wittje, R., 'Large Sound Amplification Systems in Interwar Germany: Siemens and Telefunken', *Sound and Science: Digital Histories*, https://soundandscience.de, accessed 7 November 2022

36 Humanity's Message for Whom? The Voyager Golden Record

Gambino, Megan, 'What Is on Voyager's Golden Record?' *Smithsonian Magazine*, www.smithsonianmag.com, 22 April 2012
Grebowicz, Margret, 'Why Send Whale Song into Space?' *Literary Hub*, 15 September 2017
LaFrance, Adrienne, 'Solving the Mystery of Whose Laughter Is on the Golden Record', *The Atlantic*, 30 June 2017
NASA, 'Mission Status', California Institute of Technology, https://voyager.jpl.nasa.gov, accessed 26 December 2021
—, 'The Golden Record', California Institute of Technology, https://voyager.jpl.nasa.gov, accessed 26 December 2021

Acknowledgements

Michaela Vieser and Isaac Yuen would like to thank the following people and institutions.

The Berlin Academy of Arts for their INITIAL grant, which allowed us to visit many of the locations and interviewees; the Robert Bosch Foundation for the Border Crossing Scholarship, which made a very long train journey to the Altai Mountains possible; and the Okeanos Foundation for the Sea, which led us to the whales. For the last, we would like to thank Dieter Paulmann in particular for his tireless support, the many stories shared and the mutual immersion in new worlds, as well as Jana Steingässer for her loving support in all our endeavours.

Thank you to Vera Michalski-Hoffmann, Guillaume Dollmann and the Jan Michalski Foundation for Literature in Switzerland. The essence of this book lies in our first shared listening experience among the forests around Montricher.

Thank you to Gösta Funke and Susanne Barbey for their willingness to introduce us to the moody harpsichord. To Lorenz Schreiber for his tutelage in Bach. To Ko Ishikawa for a long and sustained dialogue on classical Japanese music that awakened something profound. To Jacob Kirkegaard for an evening in an old room. To Justin Bennett for igniting the first sparks in sound and storytelling. To Stephan Malane, who prefers to play the alphorn in front of a rock face and listen to the echo. To Lucinda Newton-Dunn for hinting to the pigeon whistles in Oxford. To Will Martin, who raised new questions as a noise expert. To Matas Petrikas, who threw us into the space of sound with a text about the doorknob of the Philharmonie. To Tim Hinman for his sound advice and him and Miriam Nielsen for their friendship.

Thank you also to Baldur, Elisabeth and Charlotte, for listening.